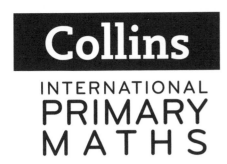

Collins
INTERNATIONAL
PRIMARY
MATHS

Workbook 3

William Collins' dream of knowledge for all began with the publication of his first book in 1819. A self-educated mill worker, he not only enriched millions of lives, but also founded a flourishing publishing house. Today, staying true to this spirit, Collins books are packed with inspiration, innovation and practical expertise. They place you at the centre of a world of possibility and give you exactly what you need to explore it.

Collins. Freedom to teach.

Published by Collins
An imprint of HarperCollins*Publishers*
The News Building
1 London Bridge Street
London
SE1 9GF

HarperCollins*Publishers*
Macken House,
39/40 Mayor Street Upper,
Dublin 1,
D01 C9W8, Ireland

Browse the complete Collins catalogue at
www.collins.co.uk

10 9 8 7

ISBN 978-0-00-836947-7

British Library Cataloguing-in-Publication Data
A catalogue record for this publication is available from the British Library.

Author: Caroline Clissold
Series editor: Peter Clarke
Publisher: Elaine Higgleton
Product developer: Holly Woolnough
Project manager: Mike Harman (Life Lines Editorial Services)
Development editor: Joan Miller
Copyeditor: Tanya Solomons
Proofreader: Catherine Dakin
Cover designer: Gordon MacGilp
Cover illustrator: Ann Paganuzzi
Typesetter: Ken Vail Graphic Design
Illustrators: Ann Paganuzzi and QBS Learning
Production controller: Lyndsey Rogers
Printed and bound in India by Replika Press Pvt. Ltd.

With thanks to the following teachers and schools for reviewing materials in development: Antara Banerjee, Calcutta International School; Hawar International School; Melissa Brobst, International School of Budapest; Rafaella Alexandrou, Pascal Primary Lefkosia; Maria Biglikoudi, Georgia Keravnou, Sotiria Leonidou and Niki Tzorzis, Pascal Primary School Lemessos; Taman Rama Intercultural School, Bali.

MIX
Paper | Supporting responsible forestry
FSC
www.fsc.org
FSC™ C007454

Contents

How to use this book

This book is used during the middle part of a lesson when it is time for you to practise the mathematical ideas you have just been taught.

- An **objective** explains what you should know, or be able to do, by the end of the lesson.

You will need
- Lists the resources you need to use to answer some of the questions.

There are two pages of practice questions for each lesson, with three different types of questions:

1 Some question numbers are written on a **circle**. These questions may be **easier**. They may also practise mathematical ideas you have learned before. These questions will help you answer the rest of the questions on the two pages.

3 Some question numbers are written on a **triangle**. These questions provide **practice** on mathematical ideas you have just been taught. They help you to understand the ideas better.

5 Some question numbers are written on a **square**. These questions are slightly more **challenging**. They make you think more deeply about the mathematical ideas.

You won't always have to answer all the questions on the two pages. Your teacher will tell you which questions to answer.

 HINT

Draw a ring around the question numbers your teacher tells you to answer.

 Questions with a star beside them require you to Think and Work Mathematically (TWM). You might want to use the TWM Star at the back of the Student's Book to help you.

Date: _____

At the bottom of the second page there is room to write the date you completed the work on these pages. If it took you longer than 1 day, write all of the dates you worked on these pages.

Self-assessment

Once you've answered the questions on the pages, think carefully about how easy or hard you find the ideas. Draw a ring around the face that describes you best.

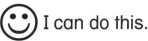 I can do this.

 I'm getting there.

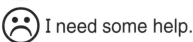 I need some help.

Number

Lesson 1: **Counting**

- Count on and count back in steps of 10 and 100

1 These patterns are counting on in 10s. Write the missing numbers.

a 14, 24, ☐ , 44, ☐ , ☐ , ☐ , 84

b 28, ☐ , ☐ , ☐ , 68, ☐ , 88

c ☐ , ☐ , 53, 63, ☐ , ☐

2 Count back in 10s from 98.

98, ☐ , ☐ , ☐ , ☐ , ☐ , ☐ , ☐ , ☐

3 Count back in 10s from 472.

⑤ 472, ☐ , ☐ , ☐ , ☐ , ☐ , ☐ , ☐ , ☐

Describe the pattern.

_____

_____

_____

4 Count on in 10s from 187.

187, ☐ , ☐ , ☐ , ☐ , ☐ , ☐ , ☐ , ☐

What do you notice about the ones? _____

What do you notice about the tens? _____

What do you notice about the hundreds? _____

5 This pattern is counting back in 100s. Write the missing numbers.

945, ☐ , ☐ , 645, ☐ , ☐ , 345, ☐ , ☐ , ☐

Number

6 Jaden counts on in 100s for four steps. He ends on 792.

What number did he start on? ☐

7 Max counts on in 10s from 85. Draw a ring around the numbers that he will **not** say.

105   147   125   231   485

8 Amira counts back in 10s for five steps. She ends on 380.

What number did she start on? ☐

9 Count back in 10s from 346.

346, ☐, ☐, ☐, ☐, ☐, ☐, ☐, ☐

What happens when you get to 306?

_____

_____

_____

10 Count on in 10s from 839.

839, ☐, ☐, ☐, ☐, ☐, ☐, ☐, ☐

What happens when you get to 899?

_____

_____

_____

11 You count back to 370 in 10s. You say eight numbers.

What was your starting number? ☐

Date: _____   ☺ 😐

Lesson 2: **Even and odd numbers**

• Recognise even and odd numbers

1 Draw a ring around the even numbers.

12   3   27   44   68   79   81   90

2 Draw a ring around the odd numbers.

25   6   34   20   37   99   21   48

3 Count on in even numbers.

12 ☐   16 ☐ ☐ ☐   24 ☐ ☐

4 Count on in odd numbers.

21 ☐   25 ☐ ☐ ☐   33 ☐ ☐

5 Sort these numbers into the Carroll diagram.

94   85   368   297   560   689   236   183

Odd numbers	Not odd numbers

6 How have the numbers been sorted? Fill in the labels on the Carroll diagram.

24	178	437	149
420	786	911	283
972	344	529	715

Number

7 Fahim thinks that 574 is an odd number because there are more odd digits than even. Is she correct? Explain your thinking.

_____

_____

8 Make a list of ten even numbers between 400 and 600.

_____

_____

9 Frankie is thinking of an odd number. The digits in his number are 3, 4 and 6. What number could he be thinking of?

There are two possibilities – write them both. ⬚ ⬚

10 Write a rule for even numbers. It must be true for all of them.

_____

_____

11 Write a rule for odd numbers. It must be true for all of them.

_____

_____

12 Jared counts back in 1s three times. He ends up on an even number with 5 hundreds and 2 tens. What number could he have started on? Write all the possibilities and show any working out.

Date: _____

9

Number

Lesson 3: **More about even and odd numbers**

• Recognise even and odd numbers

1 Draw a line to show that these even numbers can be divided into two equal groups.

Here is an example: Or you could do it this way:

a   **b**    **c**    **d**

2 Draw a line to show that these odd numbers cannot be divided into two equal groups.

a   **b**    **c**    **d**

3 Choose your own numbers to write in this table.

Even numbers	Odd numbers

Number

4 Draw a ring around the numbers that can be shared equally between 2. Are the numbers even or odd? Show this in the table.

94   59   82   125   41   96   118   17

40   91   43   162   74   89   200

Even	Odd

5 Crystal thinks that 1 cannot be an even or an odd number because it cannot be divided by 2. Is she correct? Explain your thinking.

_____

_____

6 Ahmad thinks 148 is an even number. Explain why he is correct.

_____

_____

7 Sophia thinks 167 is an odd number. Explain why she is correct.

_____

_____

8 Write a rule for even numbers.

_____

_____

What other rule can you write about even numbers?

_____

_____

9 Write a rule for odd numbers.

_____

_____

What other rule can you write about odd numbers?

_____

_____

Date: _____ 😊 😐 ☹

Number

Lesson 4: **Estimating**

- Estimate the number of objects

You will need
- exercise book

1 Draw a ring around the range that describes each estimate.

a 30–40   40–50   50–60

I estimate that there are 42 cars in the car park.

b 20–30   30–40   40–50

There are about 28 children in my class.

c 50–60   60–70   70–80

I estimate that there are 75 pencils in that pot.

2 Estimate how many apples there are. Draw a ring around the range.

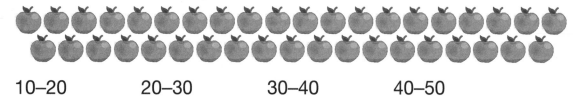

10–20   20–30   30–40   40–50

3

a Estimate how many bananas there are.

Give your answer as a range.

b Draw a ring around a group of 10 bananas. Keep making groups of 10 to help you count them.

c How many bananas are there altogether?

Number

4

a Estimate how many oranges there.

Give your answer as a range.

b Draw rings around groups of 10 oranges to help you count them.

c How many oranges are there altogether?

5 Why is it easier to estimate a range than a number?

_____

_____

_____

6 Without looking, write an estimate (as a range) for the number of pages that are in an exercise book.

a Estimate:

b Count the first 10 pages in the exercise book and hold them between your fingers.

Now you know what 10 pages look like, alter your range if you want to make it more accurate.

Refined estimate:

c Now count the pages to see what the actual answer is.

There are ___ pages in total.

Date: _____

Lesson 1: **More about estimating**

* Estimate quantities

1 Draw a ring around the range that describes each estimate.

a Susi estimates that there are 74 marbles in the pot.

50–60   70–80   90–100

b Hosea estimates that there are 96 paper clips in the box.

90–100   100–110   110–120

c Marissa estimates that her ribbon is 85 cm in length.

60–70   70–80   80–90

2 a Estimate how many fir cones there are.

b Draw a ring around a group of 10 fir cones. Keep making groups of 10 to find out how close your estimate was.

How many fir cones are there?

 3 a Estimate how many marbles there are.

Give your answer as a range.

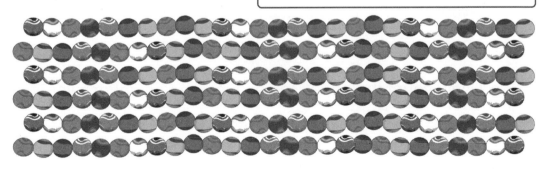

b Draw rings around groups of 10 marbles.

There are ☐ marbles.

Number

4 a Estimate how many pencils there are.

Give your answer as a range.

b Draw rings around groups of 10 pencils.

There are ☐ pencils.

5 Kayla estimated there were 687 children in her school. She gave a range of 500–600. Explain why her range is not a good one.

_____

_____

_____

6 a Write an estimate, as a range, of the number of children in your school.

b Explain why you gave this range.

_____

_____

7 Now find out how many children there are in your school.

Was your range a good one? ☐

Explain why it was or was not a good range.

_____

_____

Date: _____ ☺ 😐 ☹

Number

Lesson 2: **Counting on and back**

• Recognise patterns when counting in steps of different sizes

You will need
• coloured pencils

1 **a** Join the numbers in order, counting in 2s, to form a sequence.

2

12

8

10

6

4

14

b Draw a ring around the word that describes all the numbers in the sequence.

odd

even

2 **a** Join the numbers in order, counting in 5s, to form a sequence.

5

15

10

35

20

30

25

b Draw a ring around the word that describes all the numbers in the sequence:

odd

even

both

3 **a** Complete this sequence by counting in steps of 4.

4 ☐ 12 16 ☐ ☐ 28 ☐ 36 ☐ ☐ 48

b Draw a ring around the word that describes all the numbers in this sequence

even

odd

4 **a** Count in 2s. Put a red tick (✓) by each number you say.
Count in 3s. Put a yellow tick (✓) by each number you say.
Count in 4s. Put a blue tick (✓) by each number you say.
Count in 6s. Put a green tick (✓) by each number you say.

1	2	3	4	5	6	7	8	9	10
11	12	13	14	15	16	17	18	19	20
21	22	23	24	25	26	27	28	29	30
31	32	33	34	35	36	37	38	39	40

Number

b What do you notice when counting in 2s and 4s?

_____

c What do you notice when counting in 3s and 6s?

_____

5 For each question, colour the numbers Meera says.

a Meera counts in 3s from 27.

(39)   (31)   (43)   (45)   (37)

b She then counts back in 4s from 52.

(48)   (27)   (30)   (36)   (31)

c Meera counts on in 6s from 18.

(22)   (30)   (36)   (45)   (60)

6 a

1	2	3	4	5	6	7	8	9	10
11	12	13	14	15	16	17	18	19	20
21	22	23	24	25	26	27	28	29	30
31	32	33	34	35	36	37	38	39	40
41	42	43	44	45	46	47	48	49	50

Count in 2s from 24. Put a red line through each number.

Count in 3s from 12. Put a blue line through each number.

Count in 4s from 20. Put a yellow line through each number.

Count in 5s from 15. Put a green line through each number.

b Which numbers have three lines? [   ]

c Explain why none of the numbers have four lines.

_____

Date: _____

17

Number

Lesson 3: **Making sequences with numbers**

• Recognise and describe number sequences

1 Complete the sequence by counting in steps of 10.

18, ☐ , ☐ , ☐ , 58, ☐ , ☐ , ☐ , ☐ , 108

2 Write the numbers to show how part of the sequence is made.

13,   23,   33,   43,   53

13 + ☐   23 + ☐   33 + ☐   43 + ☐

3 Complete the sequence by counting in steps of 5.

79, 84, ☐ , ☐ , ☐ , 104, ☐ , ☐ , ☐ , 124

4 Write the numbers to show how part of the sequence is made.

62,   67,   72,   77,   82

62 + ☐   67 + ☐   72 + ☐   77 + ☐

5 Samira wrote this sequence:

145,   155,   165,   175,   185

145 + 10   155 + 10   165 + 10   175 + 10

What number will come tenth in this sequence? ☐

6 a Complete the sequence.

62, 60, ☐ , ☐ , ☐ , 52, ☐ , ☐ , ☐ , 44

b What is the rule?

_____

Number

7 **a** Complete the sequence.

310, ☐, 320, ☐, ☐, 335, ☐, ☐, ☐, 355

b What is the rule?

_____

8 **a** Complete the sequence.

570, 580, ☐, ☐, ☐, 620, ☐, ☐, 650,

b What is the rule?

_____

9 Look at your rule in **8** **b**. Use this rule to write another sequence of eight numbers.

☐, ☐, ☐, ☐, ☐, ☐, ☐, ☐

10 Ranjit thinks that the 5th step when counting in 2s from 745 is 755.

He knows this because he knows that 5 lots of 2 is 10, so he simply adds 10 to 745.

Is he correct? ☐

Why?

_____

_____

11 Ahmed thinks that the 10th step when counting back in 10s from 345 is 445. He used the sequence rule of 'add 10 to the previous number'.

Explain why he is incorrect.

_____

_____

Date: _____

Number

Lesson 4: **Making patterns with numbers**

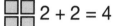

• Identify patterns in numbers

1 Complete this pattern.

▢▢ 2   ▢▢▢▢ 2 + 2 = 4   ▢▢▢▢▢▢ [ ] + [ ] + [ ] = [ ]

▢▢▢▢▢▢▢▢ [ ] + [ ] + [ ] + [ ] = [ ]

▢▢▢▢▢▢▢▢▢▢ [ ] + [ ] + [ ] + [ ] + [ ] = [ ]

▢▢▢▢▢▢▢▢▢▢▢▢ [ ] + [ ] + [ ] + [ ] + [ ] + [ ] = [ ]

2 Draw the next part of the pattern.

3 Look at this pattern.

△△△ 3 + 3 + 3 = 9 sides

△△△△ 3 + 3 + 3 + 3 = 12 sides

△△△△△ 3 + 3 + 3 + 3 + 3 = 15 sides

Draw the 4th and 5th parts of the pattern and write the addition statements.

4 Abi drew a pattern of squares that increases each time.

The first part of her pattern was made of 4 squares. 4

a Draw the next three parts of her pattern.

Number

b Write the three number statements to describe the pattern parts you have drawn.

5 Draw this pattern: 20, 15, 10, 5

6 a Draw this pattern: 3, 6, 9, 12

b What is the rule?

7 Carla wrote the number statements for her pattern.

This is the beginning of what she wrote:

6   6 + 6 = 12   6 + 6 + 6 = 18   6 + 6 + 6 + 6 = 24

The answer to her last number statement is 48.

Write the number statement to show how she made 48.

_____

8 Ali makes a number pattern. He uses squares to make his pattern.

His number pattern has 9 in it. Draw his pattern.

How many different ways can you do this?

Date: _____

21

Lesson 1: **Numerals and words (A)**

> • Read and write numbers as numerals and in words

1 Raj knows that 13 written as a word is 'thirteen', which is 1 ten and 3 ones.

Write as a numeral and draw pictures like Raj's for these numbers.

a eighteen

b fifteen

2 Draw lines to match each numeral with its written number.

89	nine hundred and twenty
276	six hundred and sixty-four
664	eighty-nine
709	two hundred and seventy-six
920	seven hundred and nine

3 Use each set of words to make two different 3-digit numbers.

  a hundred   forty   five   and   eight

  _____

  _____

  b sixty   hundred   and   seven   three

  _____

  _____

Number

Number

4 Nihal wrote some numbers in words. Some of them are incorrect. Draw a ring around the words that are spelt incorrectly. Write what Nihal should have written.

a eit hundred and fourty-nine

_____

b nin hundred and fivety-free

_____

5 Write these words as numerals.

a three hundred and fifty-six [   ]

b fifty-four [   ]

c eight hundred and ninety-nine [   ]

d one hundred and five [   ]

6 Make as many different numbers as you can, using four of these words each time. Write the numbers as numerals.

three   eight   one   fifty   seventy   hundred   and

[   ]

7 Anya is thinking of a number. Each digit is an odd number. All the digits are different. It is less than 230. What number could she be thinking of? Write all the possibilities as numerals.

[   ]

Date: _____

Number

Lesson 2: **Numerals and words (B)**

• Read and write numbers as numerals and in words

1 Draw lines to match the numerals with their hundreds.

789   4 hundreds

346   3 hundreds

183   7 hundreds

479   1 hundred

2 Draw lines to match the numerals with their tens.

567   8 tens

106   4 tens

782   6 tens

641   0 tens

3 What number does this represent?

a Write the numeral. [ ]

b Write the number in words.

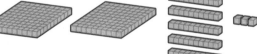

_____

4 What number does this represent?

a Write the numeral. [ ]

b Write the number in words.

_____

Number

5 Draw counters to represent the number 354.

Write the number in words.

_____

6 Complete the numbers.

a 254 _____ hundred and _____-four

b 607 six _____ _____ seven

c 498 four _____ _____ ninety-_____

d 160 _____ _____ and _____

7 Anja had these digit cards: 5 6 7

She made up six 3-digit numbers.

What were they? Write them in numerals and words.

☐ _____   ☐ _____

☐ _____   ☐ _____

☐ _____   ☐ _____

8 Samson is thinking of an odd number. The digits in his number are
5, 6 and 7. Write the possible answers as numerals.

How many possibilities are there? ☐

Date: _____

Number

Lesson 3: **Reading and writing even numbers to 1000**

• Read and write even numbers as numerals and in words

1 Underline the even numbers.

five hundred and thirty-seven   four hundred and eight

four hundred and fifty-nine   seven hundred and twelve

nine hundred and fourteen   one hundred and seventy-one

2 Underline the numbers that have an even number of tens.

three hundred and thirty-five   four hundred and forty

four hundred and sixteen   five hundred and eight

three hundred and forty-four   nine hundred and seventy-three

3 Write these even numbers in words.

a 456 _____

b 142 _____

c 740 _____

d 108 _____

e 214 _____

4 Write these even numbers as numerals.

a two hundred and sixteen ☐

b five hundred and eight ☐

c three hundred and forty-two ☐

d six hundred and eighteen ☐

e seven hundred and ninety-four ☐

Number

5 Write all the different 3-digit even numbers that you can make from these digits. Write your numbers as numerals and in words.

| 4 | 3 | 8 |

_____

_____

_____

_____

6 Here are Ben's vocabulary cards.

| fourteen | | hundred | | ninety | | and |

| six | | three | | eight |

He thinks that he can make 6 different 3-digit even numbers.

Do you agree?

Why?

_____

_____

_____

7 Grace thought that it was impossible to make an even 3-digit number with 3 in it. Is she correct? Explain your thinking. Give examples.

_____

_____

_____

Date: _____

Number

Lesson 4: **Reading and writing odd numbers to 1000**

• Read and write odd numbers as numerals and in words

1 Underline the odd numbers.

one hundred and forty-three   seven hundred and five

two hundred and sixty-seven   four hundred and nine

four hundred and fifty-two   three hundred and fifty-two

2 Underline the numbers that have an odd number of hundreds.

three hundred and nineteen   one hundred and nine

six hundred and twenty-four   eight hundred and seventy-four

two hundred and thirty-one   nine hundred and seventeen

3 Write these odd numbers in words.

a 231 _____

b 453 _____

c 579 _____

d 117 _____

e 805 _____

4 Write these odd numbers as numerals.

a five hundred and thirteen ☐

b three hundred and seven ☐

c eight hundred and forty-five ☐

d nine hundred and eleven ☐

e two hundred and nine ☐

Number

5 Write all the different 3-digit odd numbers that you can make from these digits. Write your numbers as numerals and in words.

| 7 | 5 | 6 |

[ ] _____

[ ] _____

[ ] _____

[ ] _____

6 Here are Teja's vocabulary cards.

| seventeen | | hundred | | five | | and |

| sixty | | four | | nine |

She thinks that she can make 6 different 3-digit odd numbers.

Do you agree? [ ]

Why?

_____

_____

_____

7 Manu thought that it was impossible to make an odd 3-digit number with 8 in it. Is he correct? Explain your thinking. Give examples.

_____

_____

_____

Date: _____

Number

Lesson 1: **Commutativity**

- Understand that addition can be done in any order, but subtraction cannot

1 Draw lines to match the addition statements.

10 + 40	13 + 14
24 + 3	30 + 90
50 + 28	40 + 10
8 + 7	7 + 4
14 + 13	3 + 24
90 + 30	55 + 42
42 + 55	28 + 50
4 + 7	7 + 8

2 Underline the calculation that is correct in each pair.

a 30 − 10 = 20   10 − 20 = 30

b 5 − 2 = 3   2 − 5 = 3

c 20 − 57 = 37   57 − 20 = 37

d 20 − 50 = 30   50 − 20 = 30

e 76 − 42 = 34   42 − 76 = 34

3 Write three of your own examples to show that addition can be done in any order.

_____

_____

_____

4 Write three of your own examples to show that subtraction cannot be done in any order.

_____

_____

Number

5 Abigail said, 'If I had to add 13 and 75, my calculations would be 13 + 75 or 75 + 13. I would choose the calculation 75 + 13. I think it is more efficient to add a smaller number onto a greater number.'

a Does it matter which way around she adds the numbers?

b Would you use Abigail's method? ⬜ Why?

_____

_____

6 Raj said, 'If I had to subtract 23 from 58, my calculations would be 23 − 58 or 58 − 23. Either way, I would get a difference of 35.'

Is Raj correct? ⬜ Explain your thinking.

_____

_____

7 Clara was given these numbers. 5, 14, 40, 52. She chose two numbers and wrote a pair of addition calculations.
Example: 5 + 14 = 19 and 14 + 5 = 19

a Write all the possible pairs of calculations that Clara could make.

b Explain why each pair of calculations have the same total.

_____

_____

Date: _____

Number

Lesson 2: **Complements of 100 and multiples of 100**

- Know pairs of numbers that total 100
- Add and subtract multiples of 100

You will need
- coloured pencils

1 Complete the addition calculations.

a ☐ + 20 = 100

b 70 + ☐ = 100

c 60 + ☐ = 100

d 90 + ☐ = 100

2 a Colour pairs of balloons in the same colour to show numbers that total 100.

5   15   25   35   45   55   65   75   85   95

b Choose one pair of balloons. Write the two addition facts and the two subtraction facts.

☐   ☐

3 Write two addition and subtraction facts for each pair of numbers.

a 61 and 39   ☐   ☐

b 46 and 54   ☐   ☐

c 300 and 400   ☐   ☐

Number

4 Malachi says: 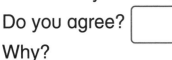 I know that 7 + 2 = 9 so, 70 + 20 = 90 and 700 + 200 = 900.

Do you agree? ⬚

Why?

_____

5 Write the missing numbers.

a 76 + ⬚ = 100

b 100 − ⬚ = 51

c ⬚ + 200 = 400

d 400 − ⬚ = 100

e 300 + ⬚ = 700

f ⬚ − 200 = 700

g 100 − ⬚ = 18

h 92 + ⬚ = 100

6 Ruth thinks that 57 + 43 = 90. Do you agree? ⬚ Why?

_____

_____

7 Jo puts eight of these balloons into four pairs that each total 1000.

Use the clues to work out the pairs. Write them on the balloons.

a The number on one of the balloons is 200 less than the number on the other one.

b The number on one of the balloons has the largest number on it.

c Both numbers begin with digits that are odd numbers.

d The number on one of the balloons is four times the number on the other one.

Date: _____

Number

Lesson 3: **Addition and subtraction of 2-digit numbers**

• Estimating, adding and subtracting 2-digit numbers

1 Work out the answer to each calculation.

a $56 + 8 =$ ▢

b $75 + 6 =$ ▢

c $83 - 6 =$ ▢

d $62 - 7 =$ ▢

e $39 + 7 =$ ▢

f $83 - 9 =$ ▢

2 Estimate the answer to each addition, writing your estimate in the bubble. Then work out the answer. Show all your working.

a $24 + 65$

b $37 + 46$

c $55 + 43$

d $68 + 87$

3 Estimate the answer to each subtraction, writing your estimate in the bubble. Then work out the answer. Show all your working.

a $82 - 56$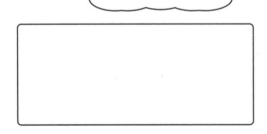

b $77 - 23$

Number

c 64 − 39

d 93 − 45

4 Arrange these digit cards to make 2-digit additions. You must use each card once in each addition.

a What is the smallest total you can make?
Show all your working.

☐☐ + ☐☐

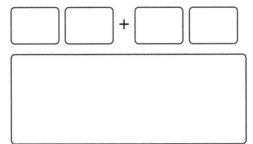

b What is the greatest total you can make?
Show all your working.

☐☐ + ☐☐

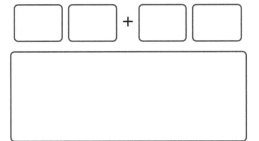

5 Arrange these digit cards to make 2-digit subtractions. You must use each card once in each subtraction.

a What is the smallest difference you can make?
Show all your working.

☐☐ − ☐☐

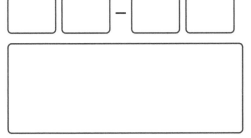

b What is the greatest difference you can make?
Show all your working.

☐☐ − ☐☐

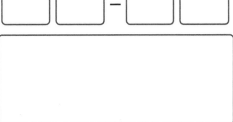

Date: _____

Number

Lesson 4: **Unknowns!**

• Use objects to represent unknown quantities

1 Additions can be written in different ways.

For example, $4 + 5 = 9$ can be written as

$4 + 5 = 9$ $5 + 4 = 9$ $9 = 5 + 4$ $9 = 4 + 5$

Write each of these additions in a different way.

a $4 + 6 = 10$ [ ]   **b** $10 = 8 + 2$ [ ]

c $2 + 3 = 5$ [ ]   **d** $8 = 4 + 4$ [ ]

e $6 + 3 = 9$ [ ]   **f** $7 = 2 + 5$ [ ]

2 Write the unknown number in each calculation.

a [ ] $+ 6 = 10$ **b** $3 +$ [ ] $= 10$

c $10 = 5 +$ [ ]   **d** $10 -$ [ ] $= 1$

e $10 -$ [ ] $= 7$ **f** $10 -$ [ ] $= 2$

3 Write the unknown number in each calculation.

a $14 +$ [ ] $= 20$ **b** $20 +$ [ ] $= 30$

c [ ] $+ 4 = 16$ **d** $18 -$ [ ] $= 13$

e $30 -$ [ ] $= 17$ **f** [ ] $- 11 = 19$

4 Make up six unknown number questions of your own.

Example: $\$15 -$ [ ] $= \$7$

What is the value of the square? [ ]

Write your questions and answers in the table.

	Questions	Answers
a		
b		
c		
d		
e		
f		

5 Choose one addition and one subtraction calculation. Solve them and then write a number story for each one.

$\boxed{} + 25 = 50$ $12 + \boxed{} = 30$

$40 - \boxed{} = 23$ $\boxed{} - 29 = 21$

a _____

_____

b _____

_____

6 **a** If you know that 5 more than an unknown number is 50, how would you find the unknown number?

_____

_____

b If you know that 50 is the total of an unknown number and 14, how would you find the unknown number?

_____

_____

Date: _____

Number

Lesson 1: **Associative property**

• Understand and use the associative property of addition

1 Draw lines to match the addition statements.

3 + 4 + 7	1 + 9 + 5
6 + 7 + 4	3 + 7 + 4
5 + 1 + 9	7 + 3 + 8
8 + 7 + 3	6 + 4 + 7

2 Explain what this calculation is showing.

$$3 + 4 + 7 = 3 + 7 + 4 \quad \underline{\hspace{6cm}}$$
$$= 10 + 4 \quad \underline{\hspace{6cm}}$$
$$= 14 \quad \underline{\hspace{6cm}}$$

3 Write three examples to show how we can add three numbers together by first making 10.

4 Write three examples to show that we can add three numbers together by first doubling one number.

5 Jack says: I find the total of 8, 7 and 3 by putting 8 in my head and then counting on 7 and then another 3. The total is 18.

Is Jack correct? ☐ What is a better way to add these numbers?

_____

_____

Number

6 Priya says:

> To add 8, 4 and 6, I add 4 and 6 first to make 10. Then I add 10 and 8. The total is 18.

Rafa says:

> If I add 8, 4 and 6, I add 8 and 4 first to make 12. Then I add 12 and 6. The total is 18.

Are Priya and Rafa both correct? ☐

Which method would you use? ☐

Explain your thinking.

_____

_____

7 Simone is given these numbers to add: 30   90   70
She makes 100 first.

  a Write Simone's method for adding the three numbers together.

  _____

  b Why is this a good method?

  _____

  _____

8 Felix is given these numbers to add together: 80   80   20
He starts with 80 and then counts on another 80 and then 20. He ends up with a total of 140.

  a Is Felix correct? ☐

  b Describe two better methods for adding the numbers.

    1 _____

    _____

    2 _____

    _____

Date: _____ ☺ 😐 ☹

Lesson 2: **Addition and subtraction of multiples of 10**

- Add and subtract multiples of 10

1 Complete the addition calculations.

  a 130 + 120 = [ ]

  b 230 + 120 = [ ]

  c 150 + 130 = [ ]

  d 250 + 130 = [ ]

2 Complete the subtraction calculations.

  a 340 – 120 = [ ]

  b 350 – 120 = [ ]

  c 380 – 230 = [ ]

  d 380 – 240 = [ ]

3 Complete the addition calculations.

  a 240 + 180 = [ ]

  b 390 + 420 = [ ]

  c 260 + 560 = [ ]

  d 530 + 280 = [ ]

  e 350 + 260 = [ ]

  f 180 + 740 = [ ]

Choose one calculation and explain how you added the numbers.

_____

_____

4 Complete the subtraction calculations.

  a 520 – 150 = [ ]

  b 610 – 420 = [ ]

  c 950 – 370 = [ ]

  d 760 – 280 = [ ]

  e 320 – 190 = [ ]

  f 440 – 170 = [ ]

Choose one calculation and explain how you did the subtraction .

_____

Number

5 Write the missing numbers

a 240 + [ ] = 360

b 360 + [ ] = 570

c 320 – [ ] = 110

d 480 – [ ] = 250

e 420 + [ ] = 840

f 570 – [ ] = 430

g 280 + [ ] = 490

h 690 – [ ] = 580

Choose one calculation and explain how you worked it out.

_____

_____

6 Saxon adds 650 and 230. He adds 65 and 23. He gets a sum

of 88. Is he correct? [ ]

Explain your thinking.

_____

_____

7 340   260   180   470   590

a Raj adds two of these numbers to make a total of 810.

Which two numbers did he choose? [ ] [ ]

Explain how you know.

_____

_____

b Sajida subtracts two of these numbers to make a difference of 330.

Which two numbers did she choose? [ ] [ ]

Explain how you know.

_____

_____

Date: _____

Lesson 3: **Addition and subtraction with 3-digit numbers**

- Estimate, add and subtract to and from 3-digit numbers

You will need

- Base 10 equipment (optional)

1 Partition these numbers into hundreds, tens and ones.
Example: $124 = 100 + 20 + 4$

a 345 _____

b 276 _____

c 387 _____

d 164 _____

e 253 _____

f 418 _____

2 Complete the calculations. If you need to, use Base 10 equipment to help you.

a $241 + 8 =$ ____

b $389 - 8 =$ ____

c $450 + 20 =$ ____

d $246 - 30 =$ ____

e $612 + 100 =$ ____

f $578 - 100 =$ ____

3 Estimate the answer to each calculation, writing your estimate in the bubble. Then work out the answer. Show all your working.

a $387 + 50$

b $295 + 400$

c $553 - 7$

d $915 - 300$

e   631 − 60

f   608 + 5

4   **a**   Draw a ring around the calculations that need regrouping when adding.

452 + 7   537 + 7   681 + 7   358 + 7

Explain how you know.

_____

_____

b   Draw a ring around the calculation with an answer that is closest to 500.

158 + 400   748 − 300   487 + 50   562 − 70

Explain how you know.

_____

_____

5   Rahmid adds 200 to a number and then adds 20. The total of Rahmid's numbers is 613. What number did he start with?

Explain how you know.

_____

_____

6   Elisa subtracts 100 from a number and then 70. She ends up with 264. What number did she start with?

Explain how you know.

_____

_____

Date: _____

Number

Lesson 4: **More unknowns!**

• Use objects to represent unknown quantities

1 Write the other addition calculation and the two subtraction calculations for each of these.

Example: $6 + 4 = 10$ | $4 + 6 = 10$ | $10 - 4 = 6$ | $10 - 6 = 4$ |

a $2 + 8 = 10$

b $12 + 8 = 20$

c $200 + 800 = 1000$

d $1 + 9 = 10$

e $11 + 9 = 20$

f $10 + 90 = 100$

2 Write the unknown number in each calculation.

a $\boxed{} + 2 = 10$

b $20 + \boxed{} = 100$

c $10 = 4 + \boxed{}$

d $100 = \boxed{} + 60$

e $10 - \boxed{} = 9$

f $\boxed{} - 90 = 10$

3 Write the unknown number in each calculation.

a $20 + \boxed{} = 70$

b $60 + \boxed{} = 100$

c $100 - \boxed{} = 90$

d $\boxed{} - 90 = 10$

e $\boxed{} + 7 = 20$

f $6 + \boxed{} = 20$

g $30 = 50 - \boxed{}$

h $10 = \boxed{} - 40$

i $20 = \boxed{} + 19$

j $100 = 50 + \boxed{}$

Make up four unknown number questions of your own.

Example: $9 + ⬭ = $15. What is the value of the oval? ▢

Write your questions and answers in the table.

	Questions	Answers
a		
b		
c		
d		

5 **a** Sunita thinks of a number. She adds 100. Her new number is 279. What number did she start with? Show your working.

_____

_____

b Bobbie thinks of a number. He subtracts 50. His new number is also 279. What number did he start with? Show your working.

_____

_____

6 Complete the calculation then write a number story.

50 = ⬭ + 24 _____

_____

7 15 more than an unknown number is 100. What is the unknown number? ▢ How do you know?

_____

_____

Date: _____

Number

Lesson 1: **Add 3-digit numbers and tens (A)**

- Estimate and add 3-digit numbers and tens

You will need
- squared paper

1 Draw lines to match each addition calculation to its representation.

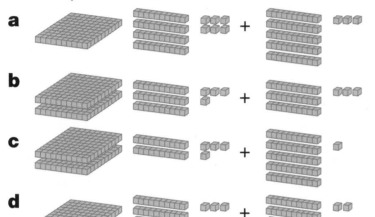

a

224 + 51

b

143 + 32

c

146 + 53

d

234 + 33

2 Estimate the answers.

a 123 + 54 ☐

b 231 + 59 ☐

c 412 + 71 ☐

d 476 + 21 ☐

e 532 + 43 ☐

f 654 + 23 ☐

3

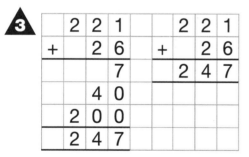

	2	2	1
+		2	6
			7
		4	0
	2	0	0
	2	4	7

	2	2	1
+		2	6
	2	4	7

Show how you would use each of these methods to find the sum of 221 and 27.

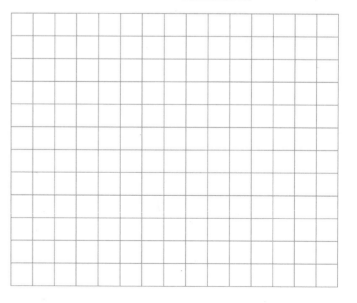

46

Number

 4 Use a **mental strategy** to add these pairs of numbers.

a 253 + 41 = [ ]

b 342 + 35 = [ ]

c 456 + 31 = [ ]

d 523 + 64 = [ ]

e 712 + 54 = [ ]

f 612 + 74 = [ ]

g Circle one calculation and explain your strategy.

_____

_____

 5 Use squared paper to work out the answer to each calculation. First **estimate** the answer, then use the **expanded written method**.

a 237 + 42 = [ ]

b 364 + 35 = [ ]

 6 Use squared paper to work out the answer to each calculation. First **estimate** the answer, then use the **formal written method**.

a 873 + 24 = [ ]

b 322 + 45 = [ ]

7 Find the sum of 345 and 24 on squared paper. Use a **mental strategy**, the **expanded written method** and the **formal written method**. Be sure to **estimate** your answer first.

8 Maisie uses the digits 1, 4, 5, 7 and 3 to make an addition.

She estimates the answer to be: 750 + 40 = 790.

a What might Maisie's numbers be? [ ] + [ ]

b Work out the answer to the calculation you have written.

[ ]

Date: _____

Number

Lesson 2: **Add 3-digit numbers and tens (B)**

* Estimate and add 3-digit numbers and tens by regrouping

You will need
* squared paper

1 Write the multiple of 10 that each number is closest to.

a 69 [ ]   **b** 58 [ ]   **c** 77 [ ]

d 89 [ ]   **e** 38 [ ]   **f** 29 [ ]

2 Answer each addition by adding the closest multiple of 10 and adjusting.

a 72 + 9 = [ ]   **b** 26 + 19 = [ ]

c 67 + 29 = [ ]   **d** 124 + 39 = [ ]

e 236 + 49 = [ ]   **f** 348 + 29 = [ ]

3

	2	2	7
+		2	6
		1	3
		4	0
	2	0	0
	2	5	3

	2	2	7
+		2	6
	2	5	3
		1	

Show how you would use each of these methods to find the sum of 227 and 27.

4 Use a **mental strategy** to add these pairs of numbers.

a 246 + 19 = [ ]   **b** 357 + 31 = [ ]

c 414 + 39 = [ ]   **d** 536 + 51 = [ ]

e 624 + 59 = [ ]   **f** 718 + 71 = [ ]

Number

g Circle one calculation and explain your strategy.

_____

_____

5 Work out the answer to each calculation. Use squared paper. First **estimate** the answer, then use the **expanded written method** or the **formal written method**.

a 328 + 48 = ⬜ **b** 239 + 55 = ⬜

c 634 + 57 = ⬜ **d** 487 + 32 = ⬜

e 529 + 64 = ⬜ **f** 864 + 29 = ⬜

6 Find the sum of 537 and 39 using squared paper. Use a **mental strategy**, the **expanded written method** and the **formal written method**. Be sure to **estimate** your answer first.

7 445 456 429 46 37 58

a Bertie adds two numbers to make the sum of 487.

Which numbers did he use? ⬜ and ⬜
Explain how you know.

_____

_____

b Which two numbers did Mercy add to make the sum of 491?

⬜ and ⬜ Explain how you know.

_____

_____

c Faisal adds the other two numbers together. What's the sum?

⬜ + ⬜ = ⬜

Date: _____

Lesson 3: **Add two 3-digit numbers (A)**

Number

• Estimate and add 3-digit numbers by regrouping

1 Use the 'making 10' strategy to answer each calculation.

Decide which number you will make into a multiple of 10.

a 47 + 6 = [　　　]　　　　**b** 58 + 7 = [　　　]

c 45 + 37 = [　　　]　　　　**d** 58 + 23 = [　　　]

2 Use the expanded written method to answer each calculation.

Example:　**a**　　　　　**b**　　　　　**c**　　　　　**d**

		2	3	7		1	2	7		2	1	7		1	4	8		2	2	6
	+	1	2	3	+	1	1	4	+	1	3	8	+	1	1	6	+	1	4	6
			1	0																
			5	0																
	+	3	0	0																
		3	6	0																

3 Estimate the answers.

a 246 + 198 [　　　]　　　**b** 212 + 129 [　　　]

c 247 + 248 [　　　]　　　**d** 336 + 149 [　　　]

4 Use the formal written method to answer each calculation.

a 326 + 248 =　　　**b** 647 + 137 =　　　**c** 426 + 355 =

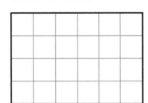　　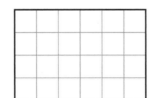　　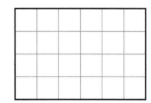

Number

d 438 + 136 =　　　　**e** 258 + 234 =　　　　**f** 549 + 138 =

5 Write the calculation in the box. Then use any method to work out the answer. Show your working on squared paper.

For example: What is 157 more than 426? | 426 + 157 = 583 |

a What is 125 more than 237?

b What is the total of 265 and 118?

c What number is 328 greater than 235?

d What is the sum of 278 and 117?

6 Steph, Robbie, Abraham and Casey had saved some money. They were each given another $128. Work out how much they each have now.

Learner	Savings	Extra money	Calculation	New total
Example	$357	$128	357 + 128 =	$485
Steph	$234			
Robbie	$246			
Abraham	$355			
Casey	$324			

Date: _____

Lesson 4: **Add two 3-digit numbers (B)**

Number

- Estimate and add 3-digit numbers by regrouping

You will need
- squared paper

1 Estimate the answers.

a 357 + 261 [ ]

b 583 + 186 [ ]

c 261 + 395 [ ]

d 472 + 447 [ ]

2 Use the expanded written method to answer each calculation.

Example: **a** **b** **c** **d**

	1	6	2
+	1	5	3
			5
	1	1	0
+	2	0	0
	3	1	5

a
	1	8	7
+	1	5	1

b
	1	8	3
+	1	4	2

c
	1	7	2
+	1	6	4

d
	2	8	3
+	1	6	2

3 Use the formal written method to answer each calculation.

a 483 + 172 =

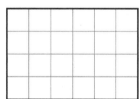

b 274 + 163 =

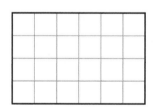

c 583 + 141 =

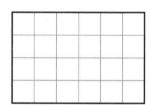

d 371 + 248 =

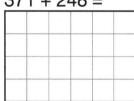

e 382 + 137 =

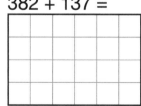

f 456 + 162 =

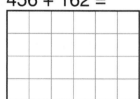

Number

 Here is a word problem.

Tommy scores 235 points on a computer game. Ritchie scores 194. How many points do they score altogether?

Altogether they score 429 points.

Write a word problem for 385 + 243 and work out the answer.

My problem: _____

_____

_____

5 Mollie starts with $6.75.

She is given another $2.84.

How much money does she

have now? [   ]

Working out

6 Shay has 252 stamps.

Marley has 184 more stamps than Shay.

a How many stamps does Marley have? [   ]

Working out

b How many stamps do they have altogether? [   ]

Working out

Date: _____

53

Number

Lesson 1: **Subtract 3-digit numbers and tens (A)**

- Estimate and subtract 3-digit numbers and tens

You will need
- squared paper

1 Draw lines to match each calculation to its difference.

a 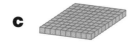 − 34

121

b − 32

114

c − 53

112

d − 74

132

2 Estimate the answers.

a 196 – 51

b 249 – 32

c 378 – 31

d 298 – 51

e 162 – 31

f 254 – 52

3

```
  300   50   6
−       10   2
  300   40   4
```

	3	5	6
−		1	2
	3	4	4

Show how you would use each of these methods to find the difference between 356 and 13.

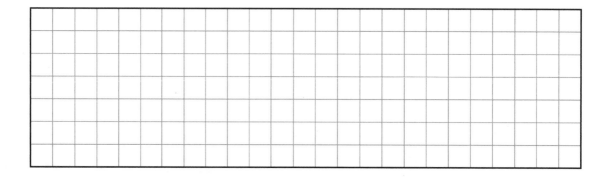

Number

 4 Use a **mental strategy** to subtract these numbers.

a 165 – 41 =

b 178 – 54 =

c 257 – 32 =

d 285 – 63 =

e 395 – 84 =

f 328 – 17 =

g Circle one calculation and explain your strategy.

_____

_____

5 Use squared paper to work out the answer to each calculation. First **estimate** the answer, then use the **expanded written method**.

a 365 – 51 =

b 198 – 62 =

c 267 – 43 =

d 468 – 27 =

6 Use squared paper to work out the answer to each calculation. First **estimate** the answer, then use the **formal written method**.

a 873 – 61 =

b 375 – 53 =

c 287 – 55 =

d 594 – 37 =

7 Find the difference between 567 and 41 using squared paper. Use a mental strategy, the expanded written method and the formal written method.

I estimate the answer to be:

The answer is:

Date: _____

Lesson 2: **Subtract 3-digit numbers and tens (B)**

Number

- Estimate and subtract 3-digit numbers and tens by regrouping

You will need
- squared paper

1 Write the multiple of 10 that each number is closest to.

a 71 [　　]

b 49 [　　]

c 17 [　　]

d 32 [　　]

e 56 [　　]

f 29 [　　]

2 Answer each subtraction by subtracting 10 and adding 1.

a 32 – 9 = [　　　　]

b 25 – 9 = [　　　　]

c 132 – 9 = [　　　　]

d 153 – 9 = [　　　　]

3

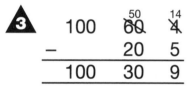

```
        50   14
100     6̶0̶   4̶
  –     20    5
100     30    9
```

	5̶	1
1̶	6̶	4
–	2	5
1	3	9

Show how you would use each of these methods to find the difference between 164 and 26.

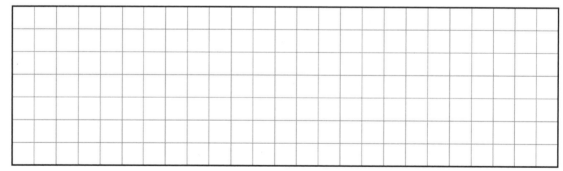

4 Use a **mental strategy** to subtract these pairs of numbers.

a 165 – 19 = [　　]

b 276 – 28 = [　　]

c 156 – 41 = [　　]

d 276 – 51 = [　　]

e 395 – 59 = [　　]

f 483 – 71 = [　　]

Number

g Circle one calculation and explain your strategy.

_____

_____

 5 Use squared paper to work out the answer to each calculation. **Estimate** first, then use the **expanded written method**.

a 145 – 27 = ☐   **b** 253 – 48 = ☐

 6 Use squared paper to work out the answer to each calculation. **Estimate** first, then use the **formal written method**.

a 265 – 48 = ☐   **b** 342 – 37 = ☐

7 Find the difference between 176 and 49 using squared paper. Use a **mental strategy**, the **expanded written method** and the **formal written method**.

8   296   384   182   48   56   75

 a Farha knows that 248 is the difference between two of the numbers.

Which numbers did he use? ☐ and ☐

Explain how you know.

_____

_____

b Which two numbers did Odin use to make a difference of 309?

☐ and ☐   Explain how you know.

_____

_____

c Moussa finds the difference of the other two numbers.

What's the calculation? ☐ – ☐ = ☐

Date: _____

Number

Lesson 3: **Subtract two 3-digit numbers (A)**

- Estimate and subtract 3-digit numbers by regrouping

You will need
- squared paper

1 Use the counting on or back strategy to find the difference.
Draw a number line to help you.

Example: 52 – 47 = 5

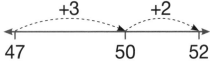

a 63 – 54 = [ ]

b 93 – 76 = [ ]

c 167 – 158 = [ ]

d 145 – 138 = [ ]

2 Use the expanded written method to answer each calculation.

a 145 – 127 =

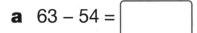

b 253 – 128 =

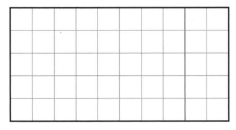

c 361 – 146 =

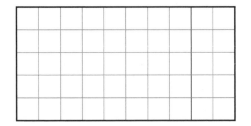

d 473 – 269 =

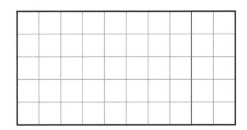

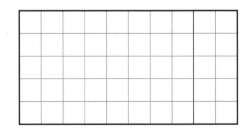

Number

 Estimate the answers.

a 246 – 128 [ ]   b 375 – 136 [ ]

c 581 – 345 [ ]   d 463 – 217 [ ]

 Use the formal written method to answer each calculation.

a 126 – 107 =   b 254 – 126 =   c 362 – 138 =

      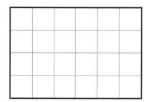

5 Write the calculation in the box. Then use any method to work out the answer. Show your working on squared paper.

For example: What is 124 less than 341? [341 – 124 = 217]

a What is 119 less than 365? [ ]

b What is the difference between 461 and 325? [ ]

c The minuend is 341. The difference is 126.

   What is the subtrahend? [ ]

d How much more is 582 than 357? [ ]

6 Bertie, Farhar, Adnan and Sabrina had saved some money. They each spent $137. Work out how much they each have now.

Learner	Savings	Spent	Calculation	New total
Example	$462	$137	462 – 137 =	$325
Bertie	$355			
Farhar	$486			
Adnan	$546			
Sabrina	$464			

Date: _____

Lesson 4: **Subtract two 3-digit numbers (B)**

Number

- Estimate and subtract 3-digit numbers by regrouping

You will need
- squared paper

1 Estimate the answers.

a 418 – 375

b 347 – 176

c 536 – 241

d 463 – 282

2 Use the expanded written method to answer each calculation.

a 238 – 155 =

b 427 – 165 =

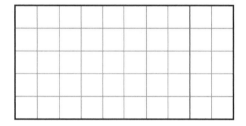
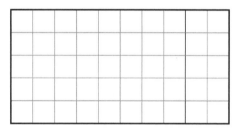

c 549 – 273 =

d 656 – 394 =

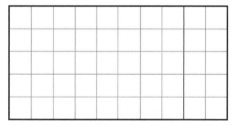
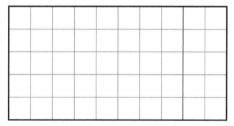

3 Use the formal written method to answer each calculation.

a 257 – 183 =   **b** 343 – 172 =   **c** 459 – 287 =

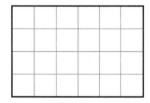
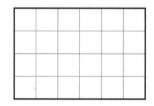
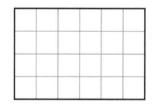

d 516 – 251 =   **e** 628 – 476 =   **f** 749 – 491 =

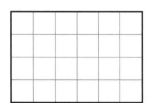
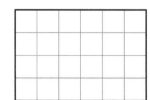
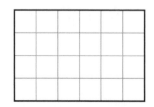

Number

4 Natasha has $236. She spends $171.

How much money does she have left? Show your working.

5 Write a word problem for 675 – 283 and work out the answer.

My problem:

_____

_____

_____

6 Samir has $8.75.

He spends $1.80 on a pen and $2.14 on an eraser.

How much money does he

have left?

Working out

7 Bea scores 345 points in a game.

Sue scores 162 fewer points than Bea.

How many points does

Sue score?

Working out

Date: _____

Number

Lesson 1: **Multiplication and division**

> • Understand the relationship between multiplication and division

1 Draw lines to match each array to a multiplication.

a

$8 \times 2 = 16$

b

$10 \times 2 = 20$

c

$5 \times 4 = 20$

2 Draw lines to match each array to a division.

a

$16 \div 8 = 2$

b

$20 \div 10 = 2$

c

$20 \div 5 = 4$

3 Draw an array for each multiplication.

a $3 \times 5 = 15$

b $5 \times 2 = 10$

c $6 \times 2 = 12$

d $4 \times 5 = 20$

Write two divisions for each multiplication.

a 8 × 5 = 40

b 5 × 2 = 10

c 6 × 5 = 30

d 5 × 10 = 50

 Complete these multiplication and division targets.

a

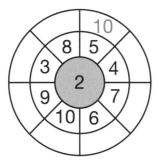

b

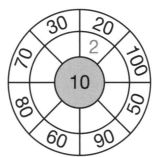

c

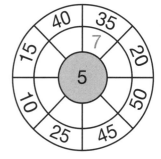

You will need to use division for these two.

6 Write two multiplications for each division.

a 20 ÷ 5 = 4

b 45 ÷ 5 = 9

c 30 ÷ 10 = 3

d 30 ÷ 5 = 6

Date: _____

Number

Number

Lesson 2: **Checking multiplication and division**

> • Use the relationship between multiplication and division

1 Write a division calculation you can use to check each multiplication.

a $5 \times 1 = 5$

b $5 \times 2 = 10$

c $3 \times 5 = 15$

d $4 \times 5 = 20$

e $6 \times 2 = 12$

f $8 \times 2 = 16$

g $6 \times 5 = 30$

h $6 \times 10 = 60$

2 Draw an array for each multiplication.

a $3 \times 2 = 6$

b $4 \times 2 = 8$

c $5 \times 2 = 10$

d $6 \times 2 = 12$

3 Write the multiplication calculation you can use to check each division.

a $20 \div 10 = 2$

b $30 \div 10 = 3$

c $40 \div 10 = 4$

d $50 \div 10 = 5$

e $60 \div 10 = 6$

f $70 \div 10 = 7$

g $80 \div 10 = 8$

h $90 \div 10 = 9$

4 a Kwame worked out that $4 \times 5 = 20$. How can he check if he is correct? Write an explanation.

_____

_____

b Melissa worked out that $40 \div 5 = 8$. How can she check if she is correct? Write an explanation.

_____

_____

 5 Draw an array that has 4 rows of 10 counters.

Now write the four calculations that you can make from this array.

 6 Show how you could check these using the inverse.

a $7 \times 10 = 70$

b $16 \div 2 = 8$

c $45 \div 5 = 9$

d $9 \times 5 = 45$

e $3 \times 10 = 30$

f $40 \div 5 = 8$

g $3 \times 5 = 15$

h $50 \div 10 = 5$

7 Reza says:

Is he correct?

Why?

I know that $6 \times 5 = 30$. I can check this by dividing 5 by 30.

_____

_____

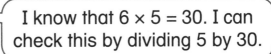

Lesson 3: **Commutativity**

• Understand that multiplication is commutative

1 Write two multiplication facts for each array.

a

b

c

d

e

2 Write a commutative fact for each of these.

a $5 \times 2 = 10$

b $6 \times 2 = 12$

c $7 \times 2 = 14$

d $8 \times 2 = 16$

3 Sami says:

I don't understand what is meant by commutative.

Write an explanation so that Sami will understand.

_____

_____

4 Pierre says:

I don't know my 9 times table, so I can't tell you what 9×2 is.

Explain what Pierre can do to work out 9×2.

_____

_____

5 Draw two arrays that each have 20 circles.

Write two multiplication facts for each one.

6 Josh's teacher says:
⊲4

> The answer to my multiplication fact is 40. What is my fact?

Write four facts that could be the answer.

7 Look at these arrays.
⊲5

a Write the multiplication facts for each array.

b How are these different from the other arrays?

_____

_____

c Draw two other arrays like this.

Date: _____

Number

Lesson 4: **Using place value to multiply**

• Understand that multiplication is distributive

1 Partition these numbers into tens and ones. Example: 13 = 10 + 3

a 12 = [　　　　]   **b** 18 = [　　　　]

c 14 = [　　　　]   **d** 15 = [　　　　]

e 16 = [　　　　]   **f** 19 = [　　　　]

2 Complete each multiplication.

Example: 11 × 2 ⟨ 10 × 2 = 20 / 1 × 2 = 2 + / 24

a 13 × 2 ⟨ 10 × 2 = [ ] / 3 × 2 = [ ] + / [ ]

b 12 × 5 ⟨ 10 × 5 = [ ] / [ ] = 10 + / [ ]

c 15 × 2 ⟨ [ ] = 20 / [ ] = 10 + / 30

d 14 × 5 ⟨ [ ] = 50 / 4 × 5 = [ ] + / [ ]

3 Answer each multiplication.

a 19 × 2 ⟨ [ ] = [ ] / [ ] = [ ] + / [ ]

b 17 × 5 ⟨ [ ] = [ ] / [ ] = [ ] + / [ ]

Number

c 16×2

d 14×5

 Use partitioning to answer each multiplication.

a $18 \times 4 =$

b $16 \times 3 =$

c $17 \times 3 =$

d $19 \times 4 =$

 This is how Billy works out 16×5. $16 + 16 + 16 + 16 + 16$

Is this a good method? Explain why.

_____

_____

 Why do 11×2 and 22×1 have the same answer?

_____

_____

_____

Write two more examples of calculations that have the same answer.

Date: _____

Number

Lesson 1: **5 and 10 times tables**

- Understand the relationship between the 5 and 10 times tables

1 Complete these times table facts.

a $5 \times 2 = \boxed{}$

b $\boxed{} \times 2 = 10$

c $5 \times \boxed{} = 15$

d $5 \times 4 = \boxed{}$

e $\boxed{} \times 5 = 25$

f $10 \times 5 = \boxed{}$

g $\boxed{} \times 7 = 70$

h $10 \times \boxed{} = 80$

i $10 \times 9 = \boxed{}$

j $10 \times \boxed{} = 100$

2 Write a division fact that goes with each of these times table facts.

a $10 \times 4 = 40$

b $5 \times 6 = 30$

c $10 \times 2 = 20$

d $5 \times 7 = 35$

e $10 \times 5 = 50$

f $5 \times 9 = 45$

3 Draw lines to match facts with the same product.

The answer when we multiply two numbers is the product.

5×2	10×5
5×4	10×3
5×8	10×1
5×6	10×4
5×10	10×2

What do you notice about the numbers in each pair?

_____

_____

4 Work out the product by multiplying by 10 and halving.

a 16 × 5 =

b 18 × 5 =

c 20 × 5 =

d 22 × 5 =

5 Show that 30 × 5 = 15 × 10.

_____

_____

6 Which of these numbers is the odd one out?

15, 25, 30, 47, 50 [ ] Explain why.

_____

_____

7 Manni says: I need to multiply 48 by 5. That's difficult!

Write an explanation to show Manni why it's not difficult.

Make sure you tell him the product.

_____

_____

Date: _____

Number

Number

Lesson 2: **2, 4 and 8 times tables**

• Understand the relationship between the 2, 4 and 8 times tables

1 Complete each 2 times table fact.

a $2 \times 8 = \boxed{}$ **b** $2 \times 4 = \boxed{}$ **c** $2 \times \boxed{} = 20$

d $2 \times \boxed{} = 6$ **e** $2 \times 5 = \boxed{}$ **f** $2 \times \boxed{} = 12$

g $2 \times \boxed{} = 18$ **h** $2 \times 7 = \boxed{}$ **i** $2 \times \boxed{} = 4$

2 Draw lines to the 2 times table fact you can double to help work out the answer to the 4 times table fact.

2×7	4×6
2×4	4×9
2×6	4×7
2×8	4×4
2×9	4×8

3 Complete each 4 times table fact.

a $4 \times 8 = \boxed{}$ **b** $4 \times 3 = \boxed{}$ **c** $40 \div 4 = \boxed{}$

d $8 \div 4 = \boxed{}$ **e** $4 \times 9 = \boxed{}$ **f** $28 \div 4 = \boxed{}$

g $16 \div 4 = \boxed{}$ **h** $4 \times 1 = \boxed{}$ **i** $4 \times 6 = \boxed{}$

4 Complete each 8 times table fact. Use doubling if you need to and show any working.

a $8 \times 4 = \boxed{}$ **b** $8 \times 6 = \boxed{}$

Number

c $8 \times 9 = \boxed{}$

d $8 \times 7 = \boxed{}$

e $8 \times 3 = \boxed{}$

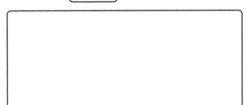

f $8 \times 8 = \boxed{}$

5 Complete each 8 times table fact.

a $32 \div 8 = \boxed{}$ **b** $16 \div 8 = \boxed{}$ **c** $80 \div 8 = \boxed{}$

d $48 \div 8 = \boxed{}$ **e** $40 \div 8 = \boxed{}$ **f** $56 \div 8 = \boxed{}$

g $24 \div 8 = \boxed{}$ **h** $8 \div 8 = \boxed{}$ **i** $72 \div 8 = \boxed{}$

j $64 \div 8 = \boxed{}$

6 Sophie says: I need to multiply 25 by 8. That's difficult!

Write an explanation to show Sophie why it's not difficult. Use the box to show how you would work it out.

Make sure you tell her the product.

_____

_____

_____

_____

_____

Date: _____

Number

Lesson 3: **Multiples**

• Recognise multiples of 2, 5 and 10 (up to 1000)

1 Write the first 10 multiples of 2.

2 Write the first 10 multiples of 5.

3 Write the first 10 multiples of 10.

4 Which is the odd one out in each set? Why?

a 2, 10, 4, 15, 12 ⬜

_____

b 5, 10, 18, 20, 15 ⬜

_____

c 30, 60, 20, 75, 90 ⬜

_____

5 Write a number that is a multiple of 2 and a multiple of 5.

⬜ Now, write another. ⬜ And another. ⬜

And another. ⬜

What do you notice?

_____

_____

Number

6 Hamid says: — I think that 2 is a multiple of 10.

Is Hamid correct? ☐

Explain your thinking.

_____

_____

7 Kati says: — I think 20 is a multiple of 2, 5 and 10.

4

Is Kati correct? ☐

Why?

_____

_____

8 Write 10 multiples of 2 between 20 and 1000.

_____

9 Mundi says: — My teacher told us that 30 is a multiple of 10. I don't understand. What is a multiple?

Write an explanation of what a multiple is to help Mundi.

_____

_____

Write three examples of multiples. Write what they are multiples of.

_____

_____

_____

Date: _____

Number

Lesson 4: **Counting in steps**

• Count in steps of the same size

1 Write the size of steps these represent.

a ●● 　 ●● 　 ●●

b ●●●● 　 ●●●● 　 ●●●●

c ●●●●● 　 ●●●●● 　 ●●●●●

d ●●●●●●●● 　 ●●●●●●●●

2 Write a times table fact to go with each step count.

a ●● 　 ●● 　 ●● 　 ●●

b ●●●●● 　 ●●●●●

c ● 　 ● 　 ● 　 ● 　 ● 　 ● 　 ● 　 ●

d ●●●● 　 ●●●● 　 ●●●●

3 Write **T** if the statement is true. Write **F** if the statement is false.

a If you step count in 2s from zero, you will say 14.

b If you step count in 5s from zero, you will say 36.

c If you step count in 8s from zero, you will say 54.

d If you step count in 10s from zero, you will say 90.

4 Mollie says: ◁— I think 4×8 is the same as step counting in 8s four times.

Is Mollie correct? ☐ Explain your thinking.

_____

_____

Number

5 Which of these numbers is the odd one out?

4, 8, 12, 14, 16 ☐ Why?

_____

_____

6 What two numbers come next in this sequence?

64, 56, 48, 40, 32, ☐ , ☐ Why?

_____

_____

7 3 + 3 + 3 + 2 + 1 can be written as the times table fact: 4 × 3.
Write each of these additions as times table facts.

a 2 + 2 + 2 + 2 + 2 + 1 + 1 ☐

Explain why you wrote this.

_____

b 4 + 4 + 4 + 4 + 3 + 1 ☐

Explain why you wrote this.

_____

c 5 + 2 + 3 + 5 + 5 + 5 + 5 + 5 ☐

Explain why you wrote this.

_____

d 8 + 8 + 6 + 8 + 8 + 2 + 8 ☐

Explain why you wrote this.

_____

e 10 + 7 + 10 + 10 + 3 ☐

Explain why you wrote this.

_____

Date: _____ 😊 😐 ☹

Lesson 1: **3 and 6 times tables**

- Understand the relationship between the 3 and 6 times tables

1 Complete each 3 times table fact.

a $3 \times 8 = \boxed{}$ **b** $3 \times 3 = \boxed{}$ **c** $3 \times \boxed{} = 27$

d $3 \times \boxed{} = 6$ **e** $3 \times 5 = \boxed{}$ **f** $3 \times \boxed{} = 12$

g $3 \times \boxed{} = 18$ **h** $3 \times 7 = \boxed{}$ **i** $3 \times \boxed{} = 30$

2 Draw lines to the 3 times table fact you can double to help work out the answer to the 6 times table fact.

3×5 6×8

3×4 6×9

3×6 6×5

3×8 6×4

3×9 6×6

3 Complete each 6 times table fact.

a $6 \times 8 = \boxed{}$ **b** $6 \times 3 = \boxed{}$ **c** $24 \div 6 = \boxed{}$

d $30 \div 6 = \boxed{}$ **e** $6 \times 9 = \boxed{}$ **f** $60 \div 6 = \boxed{}$

g $12 \div 6 = \boxed{}$ **h** $6 \times 1 = \boxed{}$ **i** $6 \times 6 = \boxed{}$

j $42 \div 6 = \boxed{}$ **k** $48 \div 6 = \boxed{}$ **l** $6 \times 7 = \boxed{}$

4 Use <, > or = to compare these times tables facts.

a $3 \times 8 \boxed{} 6 \times 4$ **b** $6 \times 9 \boxed{} 3 \times 6$

c $3 \times 3 \boxed{} 6 \times 2$ **d** $6 \times 9 \boxed{} 3 \times 10$

e $3 \times 4 \boxed{} 6 \times 2$ **f** $6 \times 4 \boxed{} 3 \times 7$

5 Shay says: _6 × 6 is greater than twice 3 × 6._

Do you agree? ☐

Explain your answer.

_____

_____

6 Look at these times table facts.

$3 \times 6 = 18$ $6 \times 3 = 18$

What is the same about them?

_____

What is different?

_____

7 My teacher says: _36 is the answer. What is my question?_

Write three possible questions.

CLUE: They must all have something to do with 3 and 6 times tables facts.

_____

_____

_____

8 All the facts for the 3 times table can be doubled to give facts for the 6 times table.

Is this always, sometimes or never true? ☐

Explain why, giving examples.

_____

_____

_____

Date: _____

Number

Number

Lesson 2: **9 times table**

• Know the 9 times table

1 Complete the number line.

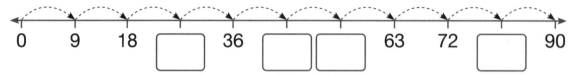

0   9   18   ☐   36   ☐   ☐   63   72   ☐   90

2 Use the number line in **1** to help you complete these.

a 9 × 1 = ☐   **b** 9 × 4 = ☐   **c** 9 × 7 = ☐

d 9 × 8 = ☐   **e** 9 × 10 = ☐   **f** 9 × 2 = ☐

3 Draw lines to match each multiplication with its product.

9 × 2 =	45
9 × 4 =	27
9 × 6 =	18
9 × 7 =	72
9 × 5 =	36
9 × 10 =	54
9 × 3 =	90
9 × 8 =	63

4 Draw lines to match each division with its quotient.

72 ÷ 9 =	1
63 ÷ 9 =	2
81 ÷ 9 =	8
27 ÷ 9 =	7
18 ÷ 9 =	4
36 ÷ 9 =	3
54 ÷ 9 =	6
9 ÷ 9 =	9

5 Show how you can use the 3 times table to work out the answer to each of these 9 times table facts.

a 9 × 8 = ☐

b 9 × 9 = ☐

Number

6 Which of these numbers is the odd one out?

27   90   35   63   45   18   36

Why is it the odd one out?

_____

_____

7 Write **T** beside the statements that are true and **F** beside the ones that are false.

a $9 \times 3 = 3 \times 3 \times 3$

b $9 \times 6 = 3 \times 6 \times 3$

c $9 \times 4 > 4 \times 9$

d $9 \times 5 < 9 \times 3 \times 2$

e $9 \times 2 + 2 = 3 \times 3 + 1$

f $9 \times 10 - 10 > 9 \times 8$

g $9 + 9 + 9 + 9 + 9 > 9 \times 6$

h $9 \times 5 = 90 \div 10$

Explain your thinking for:

c _____

_____

f _____

_____

h _____

_____

Date: _____

Number

Lesson 3: **Multiplication and division facts (1)**

• Know the 1, 2, 3, 4, 5, 6, 8, 9 and 10 times tables

1 Use this multiplication table to write ten different facts.

×	1	2	3	4	5	6	7	8	9	10
1	1	2	3	4	5	6	7	8	9	10
2	2	4	6	8	10	12	14	16	18	20
3	3	6	9	12	15	18	21	24	27	30
4	4	8	12	16	20	24	28	32	36	40
5	5	10	15	20	25	30	35	40	45	50
6	6	12	18	24	30	36	42	48	54	60
7	7	14	21	28	35	42	49	56	63	70
8	8	16	24	32	40	48	56	64	72	80
9	9	18	27	36	45	54	63	72	81	90
10	10	20	30	40	50	60	70	80	90	100

2 Write a multiplication fact that has these products. Write the corresponding division fact .

a 63

b 48

c 21

d 56

Number

3 Write the product.

  a $6 \times 4 =$ ☐   **b** $7 \times 8 =$ ☐   **c** $7 \times 10 =$ ☐

  d $4 \times 6 =$ ☐   **e** $4 \times 8 =$ ☐   **f** $10 \times 5 =$ ☐

  g $3 \times 9 =$ ☐   **h** $5 \times 7 =$ ☐   **i** $9 \times 6 =$ ☐

4 Write the quotient.

  a $54 \div 6 =$ ☐   **b** $63 \div 7 =$ ☐   **c** $24 \div 8 =$ ☐

  d $27 \div 3 =$ ☐   **e** $25 \div 5 =$ ☐   **f** $36 \div 9 =$ ☐

  g $32 \div 4 =$ ☐   **h** $60 \div 10 =$ ☐   **i** $56 \div 8 =$ ☐

5 Tabula thinks that there are three different multiplication facts that give a product of 16.

Is she correct? ☐ Why?

_____

_____

6 Write examples of a multiplication fact that has a product of 24.

☐   ☐

☐   ☐

7 Write the missing numbers in each calculation.

  a $3 \times$ ☐ $\times 2 =$ ☐   Now think of two other calculations.

  ☐

  b ☐ $\times 5 \times 2 =$ ☐   Now think of two other calculations.

  ☐

Date: _____

Number

Lesson 4: Multiplication and division facts (2)

• Know the 1, 2, 3, 4, 5, 6, 8, 9, and 10 times tables

1 Write five multiples of 2 and the multiplication facts to go with them.

Example:

| 10 | 2 × 5 = 10 |

2 Starting at zero, count on steps of 3. Write five multiples of 3 and the multiplication facts to go with them.

Example:

| 27 | 3 × 9 = 27 |

3 Starting at zero, count on in steps of 4 nine times.

What is the last number you say?

Write the two multiplication and two division facts.

4 Which is the odd one out? Why?

⑤ **a** 21, 30, 18, 25, 9

_____

b 27, 12, 36, 81, 18

_____

c 36, 60, 24, 32, 48

_____

 **5** Lotte says: | I am thinking of two numbers and their product is 30.

What two numbers could Lotte be thinking of? ☐ ☐

What other two numbers could she be thinking of? ☐ ☐

6 Yakoob says: | I think that 5 must be a quotient of 45 because 45 ends with 5.

Is this a good explanation? ☐

Explain your thinking.

_____

_____

7 Write the missing number in each trio.

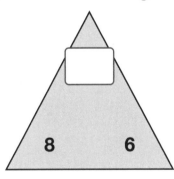 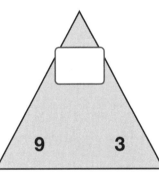 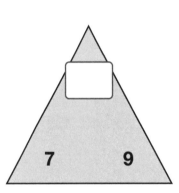

8 Georgia says: | If I multiply 9 by 8 and then halve it, I will have made a product in the 4 times table.

Is Georgia correct? ☐
Explain why.

_____

_____

Date: _____

Number

Number

Lesson 1: **Multiplying by repeated addition**

• Estimate and multiply 2-digit numbers by 2, 3, 4 and 5

1 Draw a ring around the best estimate for the product of 15 and 2:
20 40 60 80

Complete this number line to show how to work out 15 × 2.

0

2 Draw a ring around the best estimate for the product of 15 and 3:
20 40 60 80

Complete this number line to show how to work out 15 × 3.

0

3 Draw a ring around the best estimate for the product of 15 and 4:
20 40 60 80

Complete this number line to show how to work out 15 × 4.

0

4 **a** What mental calculation strategy would you use for multiplying by 2? Give an example.

_____

_____

b What mental calculation strategy would you use for multiplying by 4? Give an example.

_____

_____

Number

5 Humaira says:

> To multiply by 5, we can multiply by 10 and halve the product.

Do you agree? ☐

Why? Give an example, to show your thinking.

_____

_____

_____

6 Work out 32 × 4 in two different ways.

☐   ☐

Which way do you think is the quickest? Explain why.

_____

_____

7 Arsalan works out 23 × 4. He writes: 23 × 4 = 46

What has he done wrong?

_____

_____

What is the correct product? ☐

8 Marie solved the calculation: 43 × 2 = 806

7

Can you spot the mistake she has made?

Explain what she has done.

_____

_____

_____

Date: _____

☺ 😐 ☹

Number

Lesson 2: **Multiplying with arrays**

* Estimate and multiply 2-digit numbers by 2, 3, 4 and 5

1 Look at this array.

a What is the value of each row?

b How many rows are there?

c What is the calculation?

d What is the product?

2 Look at this array.

a What is the value of each row?

b How many rows are there?

c What is the calculation?

d What is the product?

3 Look at this array.

a What is the value of each row?

b How many rows are there?

c What is the calculation?

d What is the product?

4 Draw a place value counter array to show 32 × 3.

What is the product?

5 Here is the first row of a multiplication array.

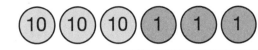

The product is 99. What is the calculation? [ ]

 6 Seth says: I can find the product of 32 and 5 by halving 32 and multiplying the answer by 10.

Show the steps that Seth will take.

 7 What possible 2-digit multiplication calculations can you make with the digits 2, 4 and 5?

Use a mental calculation strategy to find the products. Don't forget to estimate the products first!

8 Put a tick (✓) in the box if you think the calculation is correct. Put a cross (✗) if you think it is wrong.

a 42 × 3 = 126 [ ]   **b** 64 × 4 = 246 [ ]

c 58 × 5 = 290 [ ]   **d** 94 × 5 = 940 [ ]

Explain why the ones with a cross are wrong. Work out the correct products.

_____

_____

_____

_____

Date: _____

Number

Lesson 3: **Multiplying by the grid method**

Number

• Estimate and multiply 2-digit numbers by 2, 3, 4 and 5

1 Look at this array.

a   What is the value of each row?

b   How many rows are there?

c   What is the calculation?

d   What is the product?

2 Look at this grid.

×	50	3
2	100	6

```
  1 0 0
+     6
───────
  1 0 6
```

a   What is the calculation?

b   How is this the same as using counters in **1**?

_____

c   How is it different?

_____

3 Estimate the product of 21 and 5. Draw a grid and work out 21 × 5.

My estimate

4 Complete the calculations.

a | × | 20 | 4 |
    |---|----|---|
    | 2 | | |

   +

b | × | 20 | 4 |
    |---|----|---|
    | 3 | | |

   +

Number

c

×	20	4
4		

☐

+ ☐

☐

d

×	20	4
5		

☐

+ ☐

☐

5 Asima says: 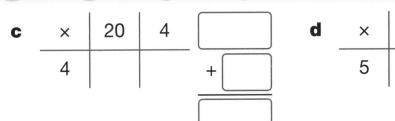 I know that $14 × 3 = 42$, so $15 × 3$ must be 45.

Is she correct? ☐ How do you know?

_____

_____

6 Draw a grid and calculate $36 × 5$. Then use a mental strategy to check your answer.

Calculate	Check

7 Fill in the missing numbers in this grid.

×	☐	5
5	400	☐

What is the product? ☐

8 Zoltan says: I know that $35 × 4$ is 140, so $35 × 5$ must be 144.

Is Zoltan correct? ☐ Why?

_____

_____

_____

Date: _____

Lesson 4: **Multiplying by partitioning**

• Estimate and multiply 2-digit numbers by 2, 3, 4 and 5

1 Partition these numbers into tens and ones.

Example: 72 = 70 + 2

a 54 = [   ]

b 89 = [   ]

c 32 = [   ]

d 28 = [   ]

2 Fill in the missing numbers.

a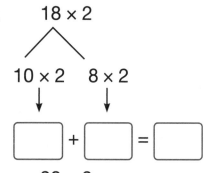

b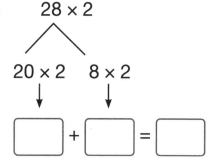

c
```
        38 × 2
        /    \
    30 × 2   8 × 2
       ↓       ↓
   [    ] + [    ] = [    ]
```

d
```
        48 × 2
        /    \
    40 × 2   8 × 2
       ↓       ↓
   [    ] + [    ] = [    ]
```

3 Use partitioning to multiply these numbers.

a 35 × 3

My estimate [   ]

b 82 × 4

My estimate [   ]

4 Jessica says: I can multiply 28 by 3 by partitioning 28 into 8 and 8 and 8 and 4. Then I can multiply each part using my times tables facts.

Can Jessica do this? [   ]

Number

Explain why. Use a diagram to help you.

5 Abraham thinks:

> If I multiply 45 by 4, all I need to do is double and double again. If I do that, I will have a product of 180.

Do you agree? ☐
Explain why.

_____

_____

Use partitioning to check Abraham's answer.

6 Maisie says:

> My teacher says I should use mental strategies to multiply by 2, 4 and 5. I don't know any.

Can you
help Maisie?

a Explain how to use a mental strategy to multiply 68 by 2.

b Explain how to use a mental strategy to multiply 68 by 4.

c Explain how to use a mental strategy to multiply 68 by 5.

Date: _____

Lesson 1: **Dividing using known facts**

Number

> • Estimate and divide 2-digit numbers by 2, 3, 4 and 5

1 Use multiplication facts to work these out.

a $24 \div 3 =$ ☐   **b** $32 \div 4 =$ ☐   **c** $16 \div 2 =$ ☐

d $40 \div 5 =$ ☐   **e** $18 \div 2 =$ ☐   **f** $27 \div 3 =$ ☐

g $30 \div 5 =$ ☐   **h** $36 \div 4 =$ ☐   **i** $18 \div 3 =$ ☐

2 What division calculations are shown here?

a ☐

3   6   9   12   15

b ☐

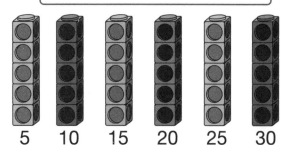

5   10   15   20   25   30

c ☐

4   8   12   16   20

d ☐

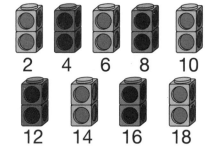

2   4   6   8   10

12   14   16   18

3 **a** What mental strategy would you use for dividing by 2?
Give an example.

_____

b What mental strategy would you use for dividing by 5?
Give an example.

_____

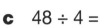
Number

4 Divide these numbers by halving and halving again.

a 24 ÷ 4 = ☐

b 44 ÷ 4 = ☐

c 48 ÷ 4 = ☐

d 56 ÷ 4 = ☐

e 88 ÷ 4 = ☐

f 72 ÷ 4 = ☐

5 Attiqa says:

> I can find the quotient of any number when dividing by 5, by halving and dividing by 10. Look at my example, 80 ÷ 5. Halve it to make 40, then divide by 10 to give a quotient of 4.

Do you agree with Attiqa? ☐ Explain your thinking.

_____

_____

6 Bernie wants to divide 84 by 4. This is what he writes. 84 ÷ 4 = 42
What has he done wrong?

_____

_____

What is the correct answer? ☐

7 Tooba has completed a calculation. 48 ÷ 2 = 204
Can you spot the mistake she has made? ☐
Explain what she has done.

_____

_____

Date: _____   95

Lesson 2: **Dividing by partitioning**

- Estimate and divide 2-digit numbers by 2, 3, 4 and 5

Number

1 Partition 46 into tens and another number.

Do it in four different ways.

(10) (10) (10) (10) (1) (1) (1) (1) (1) (1)

a ☐ and ☐   **b** ☐ and ☐

c ☐ and ☐   **d** ☐ and ☐

2 These numbers have been partitioned.

Write what the whole numbers are.

a 20 and 15 ☐   **b** 30 and 15 ☐   **c** 40 and 15 ☐

d 50 and 15 ☐   **e** 60 and 15 ☐   **f** 80 and 15 ☐

3 Draw place value counters to show how to partition 56.

Example: 50 and 6

(10) (10) (10) (10) (10) (1) (1) (1) (1) (1) (1)

Show four more ways.

☐

☐

☐

☐

4 Partition each number, then work out the answer.

a
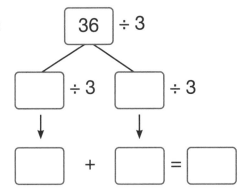

b
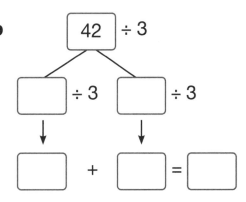

c
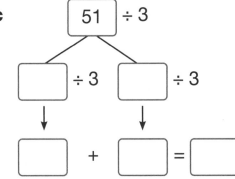

d
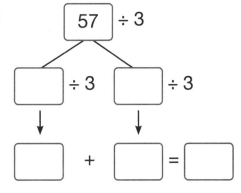

5 Show that 92 ÷ 4 is 23.

6 Write four different calculations that each have a quotient of 4.
The dividend needs to be greater than 50.

What do they all have in common?

_____

How can this help you to find others?

_____

Date: _____

Number

Lesson 3: **Dividing with arrays**

• Estimate and divide 2-digit numbers by 2, 3, 4 and 5

1 What numbers do these place value counters show?

a (10) (10) (1) (1) (1) ☐

b (10) (10) (10) (1) (1) ☐

c What is the same about these numbers?

_____

d What is different?

_____

2 Look at this grid.

10s	**1s**
(10) (10) (10) (10)	(1) (1)

Draw the groups we make if we divide this number by 2.

What is the answer? ☐

3 Write the steps and draw place value counters to divide 48 by 3.

Estimate the quotient first. ☐

Number

4 Daisy says:

Is Daisy correct? ☐

> If I divide 48 by 2, the answer will be 6 because 4 divided by 2 is 2 and 8 divided by 2 is 4. I then add 2 and 4, which is 6.

Why? _____

_____

5 Any number that ends with 5 can be divided by 5 to give a whole number.

Is this always, sometimes or never true? ☐

Why? _____

_____

6 Work out 96 ÷ 4.

Use the grid to help you.

What is the answer? ☐

Explain how you can use a mental calculation to check this division.

10s	1s
10 10 10	1 1
10 10 10	1 1
10 10 10	1 1

_____

_____

7 **a** Use the digits 2, 4 and 8 once each to make up the most difficult division that you can.

☐

Why is it the hardest?

_____

b Use the same digits to make up the easiest division that you can. Why is it the easiest?

☐

_____

Date: _____

Lesson 4: **Division with remainders**

• Estimate and divide 2-digit numbers by 2, 3, 4 and 5

1 Look at this grid.

10s	1s
10 10 10	1 1 1
	1 1 1

a Draw the groups we make if we divide this number by 3.

What is the quotient? ☐

b Write the quotient in the correct place on the calculation.

3) 3 6

2 Look at this grid.

10s	1s
10 10 10	1 1 1
10 10 10	1 1 1

a Draw the groups we make if we divide this number by 2.

What is the quotient? ☐

b Write the quotient in the correct place on the calculation.

2) 6 6

3 Draw lines to match each quotient to a calculation.

a 3) 6 5   24 remainder 1

b 3) 6 7   22 remainder 2

c 3) 6 8   21 remainder 2

d 3) 7 3   24 remainder 2

e 3) 7 4   22 remainder 1

Use the written method to answer each division.

a
$$4\overline{)5\ 7}$$

b
$$5\overline{)6\ 8}$$

c
$$3\overline{)7\ 6}$$

d
$$2\overline{)7\ 5}$$

e
$$4\overline{)9\ 7}$$

f
$$5\overline{)8\ 3}$$

5 Adisa answers a division calculation. This is what he does.

$$\begin{array}{r} 2\ \ 1 \\ 4\overline{)9\ \ 4} \end{array}$$

Explain what he has done wrong.

_____

_____

Show what he should have done.

 Jayde has 59 stickers. She sorts them into equal groups. She has some stickers remaining. How many stickers could be in each group and how many stickers would be remaining?

_____

_____

_____

_____

Date: _____

Number

Lesson 1: **Writing money**

• Write money with the decimal point

1 How many cents are there in one dollar? ⬜

2 How much money is this? Write the amount in cents.

1 1 1 1 1 1 1 1 1 1

1 1 1 1 1 1 1 1 1 1

10 10 10 10 ⬜

3 Ramjit has $1.12 in his pocket.

He has a dollar note and some coins.

Write two different ways that he might have the coins.

⬜ ⬜

4 How much money is this? Write the amount in dollars $ and cents.

10 10 10 1 1 1 1 1 ⬜

5 Libby has $5 in notes and 132 cents.

How much money does she have altogether? ⬜

6 Write each amount in dollars $ and cents.

a 3 dollars and 50 cents ⬜

b 9 dollars and 25 cents ⬜

c 10 dollars and 6 cents ⬜

d 21 dollars and 11 cents ⬜

7 What notes and coins can you use to pay for each item?

1 Write three different combinations of notes and coins for each item.

a

$5.00

b

$10.00

c

$20.00

8 a Write each amount of money in dollars $ and cents.

i

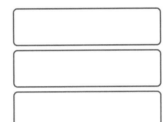

ii

b What is the difference in value between the two amounts?

Number

Lesson 2: **Finding totals**

• Add amounts of money to find totals

You will need

• selection of 1c, 5c, 10c, 25c and 50c coins

1 **a** Anji buys a pen and some stickers.

$0.90c $0.80c

Use coins to make the two amounts. Find the total.

b Draw the total. Use as few notes and coins as possible.

2 **a** Look at the prices in **1**. Paul buys three sets of stickers. Use coins to make the amounts. Find the total.

b Draw the total. Use as few notes and coins as possible.

3 Miriam buys one adult and one children's ticket. How much does Miriam spend?

Show your working.

Tickets:

Adult: $16

Children: $9.50

4 a India buys an apple, a banana and a pear.

$0.60 $0.80 $1.20 $1.25

How much does she spend? ☐

Show your working.

☐

b How can India pay for her fruit with the fewest notes and coins? Draw them in the box.

☐

5 a Look at the prices of fruit in **4**. Yukesh buys 2 oranges and a banana. How much does he spend? ☐

Show your working.

☐

b How can Yukush pay for his fruit with the fewest notes and coins? Draw them in the box.

☐

Date: _____

Number

Lesson 3: **Finding change**

- Subtract amounts of money to find change

You will need
- selection of 1c, 5c, 10c, 25c and 50c coins (optional)

1 a Cassie buys one set of stickers and gives the seller £2.

$0.90c

$0.80c

How much change does she get?

b Show your working.

2 a Look at the prices in **1**. Sam buys two pens and gives the seller $2.
How much change does he get?

b Show your working.

3 a Samir has $20. He buys a football.

$17.00

$13.00

$3.50

$7.50

How much change does he get? Show your working.

b How much change will he get from $10 if he buys a racing car?

Show your working.

c What if he buys a skipping rope and gives the seller $5?

Show your working.

d What if he buys a remote control car and gives the seller $50?

Show your working.

4 a Ibrahim has saved $15. He spends $14.15 on a chess set.

How much does he have left?

b Show your working.

5 Dalia has $20. She wants to buy a doll for $13 and a game for $7.40.

a Does she have enough money?

b Explain your answer.

Date: _____

Lesson 4: **Solving problems with money**

- Solve problems with money

1 **a** Kamla buys 3 kg of potatoes.

Each kilogram costs $3.

How much does she spend?

b Kamla also buys $\frac{1}{2}$ kg of bananas.

Each kilogram costs $4.

How much does she spend on

bananas?

Show your working.

2 Cinema tickets cost $6 each. Children's tickets are half price.

a How much will it cost a family of 2 adults and

3 children to go to the cinema?

Show your working.

b How much will it cost a group of 5 children and 3 adults to go to

the cinema?

Show your working.

c Popcorn costs $1.50 for a small carton. A large carton costs twice as much. A group of 7 children and 3 adults each buy a cinema ticket. The group also buy 6 small cartons of popcorn and 1 large carton.

$1.50

How much does the group spend, in total?

Show your working.

3 An adult's return train ticket costs $15. A child's return train ticket costs $10. A one-way ticket is half the price of a return ticket.

a 2 adults buy return tickets and another adult buys a one-way ticket.

What is the total cost?

Show your working.

b A family of 4 buy 2 adult one-way tickets and 2 child one-way tickets.

What is the total cost? Show your working.

c A family of 5 buy 2 adult return tickets and 3 child return tickets. They pay with a $100 note. How much change will they get?

Show your working.

Date: _____

109

Number

Lesson 1: **Understanding place value (A)**

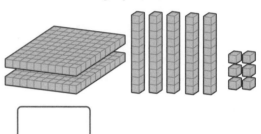

- Understand the value of each digit in a 3-digit number
- Compose and decompose 3-digit numbers, using hundreds, tens and ones

1 Write the 3-digit number made by the Base 10 equipment.

a

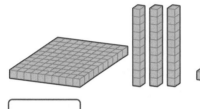

b

2 Draw Base 10 equipment to show each 3-digit number.

a 351

b 463

3 Look at each target board. What number is being shown?

a

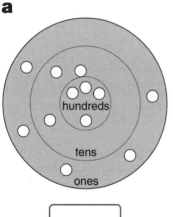

b

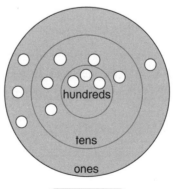

c

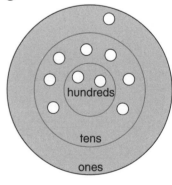

4 Write the value of each digit in these numbers.

a 647 = ⬚ hundreds + ⬚ tens + ⬚ ones

b 357 = ⬚ hundreds + ⬚ tens + ⬚ ones

c nine hundred and forty-nine =

⬚ hundreds + ⬚ tens + ⬚ ones

d seven hundred and thirty-three =

⬚ hundreds + ⬚ tens + ⬚ ones

5 Amina says:

> I can use the digits 3, 4 and 5 to make 345. I simply write the digits side by side and that's what I get.

Is this a good explanation of how to make a 3-digit number? Why?

_____

_____

_____

6 You have four digit cards. You can only use each one once in the same number.

a What is the smallest 3-digit number you can make? ⬚

b What is the largest 3-digit number you can make? ⬚

c Make all the 3-digit numbers you can with an odd number of tens.

_____

_____

d Make a 3-digit number in which the hundreds digit is worth twice as much as the tens digit. ⬚

Date: _____

Number

Lesson 2: **Understanding place value (B)**

• Understand the value of each digit in a 3-digit number

1 What numbers do these abacus show?

a  □ b  □ c  □

d  □ e  □ f  □

2 What is the value of each digit in these numbers?

a 348 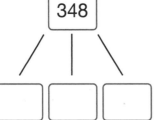 b 273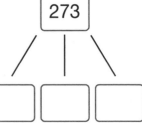

3 Write each number that the place value counters show.
Then draw a line from each number to the description
that matches.

a (100)(100)(100)(1)(1)(1) □ | The smallest number.

b (100)(100)(10)(1)(1) □ | The number with 4 tens.

c (100)(100)(100)(10)(10)(1)(1)(1)(1) □ | The odd number.

d (100)(10)(10)(10)(10)(1)(1)(1)(1) □ | The largest number.

e (100)(10)(10)(1)(1)(1)(1) □ | The number with 2 ones.

4 Write the value of each digit in these numbers.

a 532 = ☐ + ☐ + ☐   **b** 245 = ☐ + ☐ + ☐

c 701 = ☐ + ☐ + ☐   **d** 386 = ☐ + ☐ + ☐

e three hundred and fifty-nine = ☐ + ☐ + ☐

f four hundred and sixty = ☐ + ☐ + ☐

g Which of the numbers in **a** to **f** has no tens? ☐

5 Sam uses the digits 2, 5 and 7 to make some 3-digit numbers.
He makes 257, 572 and 725.

Has he made all the different numbers possible? ☐

If not, write any other numbers he can make.

_____

_____

_____

6 You have only 9 beads.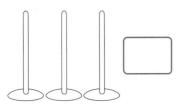

Draw 9 beads on each abacus to show each number.

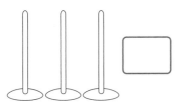   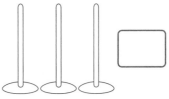

a The largest 3-digit number.   **b** The smallest 3-digit number.

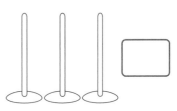

c A 3-digit number with   **d** A 3-digit number with the
4 hundreds.   same number of tens
                                      as ones.

Date: _____

Number

Lesson 3: **Regrouping**

- Understand that numbers can be regrouped in different ways

You will need
- place value counters or Base 10 equipment

1 Draw a ring around the number in which…

a	…the 8 is worth 8 tens	389	893	938
b	…the 3 is worth 3 hundreds	543	435	354
c	…the 1 is worth 1 one	821	812	182
d	…the 0 is worth 0 tens	930	390	309
e	…the 2 is worth 20	642	264	426
f	…the 6 is worth 600	613	361	136

2 Regroup these numbers into tens and ones.

a 145 [ ] tens [ ] ones   **b** 245 [ ] tens [ ] ones

c 345 [ ] tens [ ] ones   **d** 445 [ ] tens [ ] ones

3 Write the number that you could make from these place value cards.

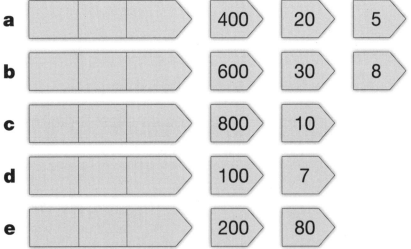

a [ ]   400 20 5

b [ ]   600 30 8

c [ ]   800 10

d [ ]   100 7

e [ ]   200 80

f [ ]   500 4 90

g [ ]   9 300 50

Number

4 Maddie regrouped a number. She made 2 hundreds,

24 tens and 13 ones. What was her number? [ ]

5 Aki says: I can regroup 546 into 5 hundreds and 46 ones.

Is Aki correct? [ ] Why? _____

_____

6 How is 125 represented?

a (100)(10)(10)(1)(1)(1)(1)(1)

[ ] , [ ] and [ ]

b (10)(10)(10)(10)(10)(10)(1)(1)(1)(1)(1)
(10)(10)(10)(10)(10)(10) [ ] and [ ]

c (10)(10)(10)(10)(1)(1)(1)(1)(1)
(10)(10)(10)(10)(1)(1)(1)(1)(1) [ ]
(10)(10)(10)(1)(1)(1)(1)(1) and [ ]

7 How else could you regroup 125? Use Base 10 equipment or place value counters to regroup 125 in other ways.

[ ]

8 Write 6 ways to regroup 246. [ ] , [ ] ,

[ ] , [ ] , [ ] , [ ]

Date: _____ 😊 😐 ☹

115

Lesson 4: **Comparing numbers**

• Compare 3-digit numbers

1 Complete each sentence to show what the symbol means.

   a The < symbol means 'is _____ than'.

   b The > symbol means 'is _____ than'.

2 Mark each comparison with a tick (✓) or a cross (✗) to show whether it is right or wrong.

   a 84 > 74 ☐   **b** 37 > 39 ☐

   c forty-nine < 67 ☐   **d** 82 < 81 ☐

   e 95 > eighty-five ☐   **f** sixty < 82 ☐

3 Write the correct symbol (> or <) between each pair of numbers.

   a 145 ☐ 154   **b** 734 ☐ 347

   c 628 ☐ 286   **d** 289 ☐ 298

   e 703 ☐ 730   **f** 294 ☐ 249

   g 101 ☐ 110   **h** 487 ☐ 478

   i 529 ☐ 295   **j** 422 ☐ 242

4 Shona makes these statements comparing 3-digit numbers. Write out her comparisons, using numerals and symbols.

   a Three hundred and twelve is more than three hundred and two.

   ☐

   b Seven hundred and eight is less than eight hundred and seventy.

   ☐

Number

c Five hundred and forty-four is more than four hundred and fifty-nine.

 5 Write a 3-digit number that makes each comparison true.

a 295 < ☐

b 402 > ☐

c 772 > ☐

d 266 < ☐

e 834 > ☐

f 333 > ☐

 6 Write a number that lies between each pair of numbers.

a 348 ☐ 386

b 579 ☐ 597

c 433 ☐ 455

d 110 ☐ 144

7 Shona and Kyle are playing a game.
Kyle makes a number out of three digit cards and writes a symbol. Shona then has to rearrange Kyle's cards to make a different number that makes the comparison true.
Can you help her? Remember that each statement must be true.

Example: 4 5 9 < 9 5 4

a 6 2 5 < ☐ ☐ ☐

b 1 9 2 > ☐ ☐ ☐

c 8 3 4 < ☐ ☐ ☐

d 5 0 2 < ☐ ☐ ☐

e 9 2 5 > ☐ ☐ ☐ or ☐ ☐ ☐

f 3 1 8 < ☐ ☐ ☐ or ☐ ☐ ☐

Date: _____

Lesson 1: **Ordering numbers**

- Order 3-digit numbers

1 Draw a ring around the largest number in each set. Draw a square around the smallest number in each set.

a 38	49	18	48	39
b 25	15	19	35	20
c 93	99	95	97	98
d 57	50	75	77	55
e 49	52	60	57	59

2 **a** Write these numbers in order, from smallest to largest.

346   146   746

b Write these numbers in order, from largest to smallest.

821   321   621

3 These tickets show the order people need to be seen in a doctor's surgery.

Write each set of tickets in order, starting with the smallest.

a

b

c 26   23   75   59

d 384   834   348   438

e 619   691   196   916

f 527   275   725   257

Number

4 In each set of numbers, one number is not in the correct position. Draw a ring around that number then order the numbers correctly.

a 832, 853, 835, 849, 852

b 409, 410, 437, 421, 439

c 571, 517, 538, 565, 569

5

Adil: To compare 3-digit numbers you compare the hundreds digits first, then the tens, then the ones.

Josh: That's not right! You always need to start with the ones.

Who is right and why?

_____

_____

6 Complete each set so that the 3-digit numbers are in ascending order. You can only use each digit once in each set of numbers.

| 0 | 1 | 2 | 3 | 4 |
| 5 | 6 | 7 | 8 | 9 |

a 5 ☐ ☐ , 5 ☐ ☐ , 5 **7** **3** , 5 ☐ ☐ , 5 ☐ ☐

b 2 ☐ ☐ , 2 **6** **4** , 2 ☐ ☐ , 2 ☐ ☐ , 2 ☐ ☐

c 7 ☐ ☐ , 7 ☐ ☐ , 7 ☐ ☐ , 7 ☐ ☐ , 7 **6** **8**

7 621   912   159   783   Cassy said:

Is Cassy correct?

Cassy: These numbers are in ascending order, because to order you add the digits of each number together and then order the totals.

☐

Why?

_____

_____

☺ 😐 ☹

Date: _____

Lesson 2: **Multiplying by 10**

• Use place value to multiply 2-digit numbers by 10

1 Draw lines to match each calculation with its answer.

8 × 10　　　　　　　　　　　　　　　　　　　　　　12 × 10

　　　　　　　　　　40　　　120

150

4 × 10　　　　　　　　　　　　　　　　　　　　　　15 × 10

　　　　80　　　160　　　100

10 × 10　　　　　　　　　　　　　　　　　　　　　16 × 10

2 Show how each digit shifts to the left when it is multiplied by 10.

Example:
39 × 10

100s	10s	1s
	3	9

× 10 →

100s	10s	1s
3	9	0

a 84 × 10

100s	10s	1s

× 10 →

100s	10s	1s

b 96 × 10

100s	10s	1s

× 10 →

100s	10s	1s

c 50 × 10

100s	10s	1s

× 10 →

100s	10s	1s

d 13 × 10

100s	10s	1s

× 10 →

100s	10s	1s

3 Complete these calculations.

a 64 × 10 = ☐　　**b** 91 × 10 = ☐　　**c** 17 × 10 = ☐

d 28 × 10 = ☐　　**e** 35 × 10 = ☐　　**f** 14 × 10 = ☐

4 This machine multiplies numbers by 10. Complete the tables.

IN
47
88
19
32

× 10

OUT
750
420
200

5 When you multiply by 10, the answer is always 10 times greater than the starting number.
Is this always, sometimes or never true? [   ] Why?

_____

_____

6 Peter thinks that when you multiply by 10 you just add a zero. Explain why he is incorrect.

_____

_____

7 George uses four different digit cards to make two different 2-digit numbers. He then multiplies each number by 10. His answers are 730 and 210.

a What are his digit cards? [   ] [   ] [   ] [   ]

b Write four more × 10 calculations with George's digits.

[   ]   [   ]

[   ]   [   ]

Date: _____   ☺ 😐 ☹

121

Number

Lesson 3: **Rounding to the nearest 10**

• Round 3-digit numbers to the nearest 10

You will need
• coloured pencil

1 a Colour the numbers that round to 70.

(64)   (68)   (71)   (69)   (73)   (76)

b Colour the numbers that round to 30.

(31)   (29)   (24)   (37)   (25)   (38)

c Colour the numbers that round to 90.

(95)   (87)   (85)   (93)   (91)   (82)

2 Round each number to the nearest 10.

a 234 ▢   **b** 219 ▢   **c** 263 ▢

d 278 ▢   **e** 292 ▢   **f** 287 ▢

3 Estimate the answer of each calculation by rounding the numbers to the nearest 10.

a 147 + 231

b 469 − 122

c 123 + 265

d 742 − 138

e 628 − 219

f 311 + 425

Number

4 Max rounded the number 345 to 340.

Is Max correct? ☐

Why?

_____

_____

5 Toby gave an estimate for 234 + 126 by rounding to the nearest 10. His estimate was 300.

Do you agree? ☐

Why?

_____

_____

6 Aki says:

> My teacher says that we need to remember this rhyme for rounding numbers up: 'five and above give them a shove'.

Is this a good thing to tell Aki? ☐

Why?

_____

_____

7 Ellison and Duke estimated the sum of 456 and 315. They rounded both numbers to the nearest 10. Ellison's estimate was 770. Duke's estimate was 780.

Whose estimate do you agree with? ☐

Why?

_____

_____

Date: _____

Number

Lesson 4: **Rounding to the nearest 100**

- Round 3-digit numbers to the nearest 100

1 Round each number to the nearest 100.

a 460 ☐ b 190 ☐ c 320 ☐

d 810 ☐ e 730 ☐ f 650 ☐

g 280 ☐ h 940 ☐ i 870 ☐

2 a Write five 3-digit numbers that round down to the nearest 100.

☐ ☐ ☐ ☐ ☐

b Write five 3-digit numbers that rounded up to the nearest 100.

☐ ☐ ☐ ☐ ☐

3 Mark each number on the number line. Show whether you would round it **up** or **down** to the nearest 100.

a 858 800 — 900

b 520 500 — 600

c 489 400 — 500

d 150 100 — 200

e 714 700 — 800

f 349 300 — 400

4 Estimate the answer of each calculation by rounding the numbers to the nearest 100.

a 289 + 154 ☐ b 621 − 378 ☐

c 193 + 227 [ ]   **d** 545 + 302 [ ]

e 685 + 350 [ ]   **f** 387 + 512 [ ]

 5 Round each number to the nearest 10 and 100.

a 341 [ ] [ ]   **b** 756 [ ] [ ]

c 901 [ ] [ ]   **d** 678 [ ] [ ]

e 148 [ ] [ ]   **f** 324 [ ] [ ]

6 Freddie and Flo made an estimate for the sum of 245 and 336.
Freddie rounded his numbers to the nearest 100. Flo rounded
her numbers to the nearest 10. Freddie's estimate was 500.
Flo's estimate was 580.

Whose estimate is better? [ ]

Why?

_____

_____

7 Keisha says:

If I add the digits of my mystery number,
they make 14. If I round my mystery number
to the nearest 100, it rounds to 500.

What could Keisha's number be? [ ]

Invent a similar puzzle of your own for a friend to try.

[ ]

Date: _____

Number

Lesson 1: **Equal parts: quantities**

• Understand and explain that fractions are one or more equal parts and all the parts, taken together, equal one whole

1 The biscuits in each of these jars are shared between two people. How many does each person get?

a **10** [ ] biscuits each   b **14** [ ] biscuits each

c **20** [ ] biscuits each   d **12** [ ] biscuits each

2 Draw lines to match each number with the number that is half of it.

16		3
22		8
6		9
18		11

3 Label each part with its fraction.

a
1 whole

b
1 whole

c
1 whole

d
1 whole

4 Complete the bars to show the values of the fractions.

a
16

b
20

c
12

d
30

5 Write the fraction of each shaded part of these diagrams.

a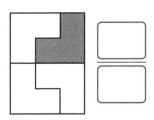

b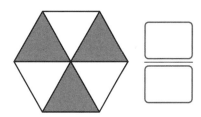

c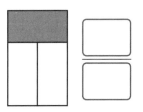

d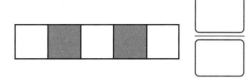

e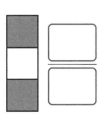

f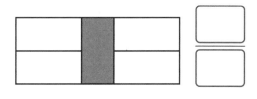

6 Draw a ring around the fraction that is the odd one out.

$\frac{1}{4}$ $\frac{1}{5}$ $\frac{7}{10}$ $\frac{1}{10}$ $\frac{1}{2}$

Explain your choice.

_____

_____

7 This is $\frac{1}{5}$ of a bag of marbles.

How many marbles are in a full bag? ☐

Explain how you know.

_____

_____

Date: _____

127

Lesson 2: **Equal parts: shapes**

- Understand that fractions are several equal parts of a shape

1 Colour each shape to show the fraction.

a 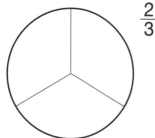 $\dfrac{2}{3}$

b 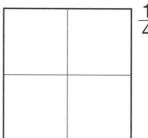 $\dfrac{1}{4}$

c 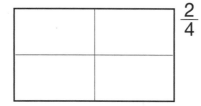 $\dfrac{2}{4}$

d 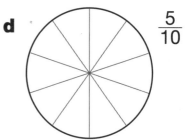 $\dfrac{5}{10}$

e 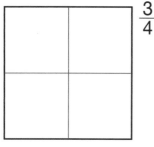 $\dfrac{3}{4}$

f 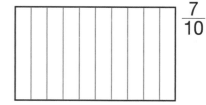 $\dfrac{7}{10}$

2 Two of the fractions you coloured in **1** are equal to $\dfrac{1}{2}$.

Which ones? ☐ and ☐

3 What fraction of each shape is shaded?

a 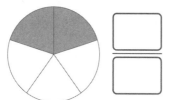 ☐ ☐

b  ☐ ☐

c 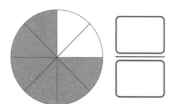 ☐ ☐

d 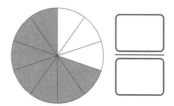 ☐ ☐

4 What fraction of each shape is shaded?

a

b

c

d

e

f

g

h

5 Which of the shapes in **4** have half shaded?

Which ones? [ ] and [ ]

6 Split each rectangle into equal parts and shade it to show a fraction.

The fractions I have made are $\frac{\square}{\square}$, $\frac{\square}{\square}$ and $\frac{\square}{\square}$.

7 Mara says this shape is divided into four equal parts. Do you agree with Mara? [ ] Explain why.

_____

_____

Date: _____

Number

Lesson 3: **Same fraction, different whole**

• Understand the relationship between 'whole' and 'parts'

1 Colour each shape to show the fraction.

a $\frac{1}{2}$ b $\frac{3}{4}$ c $\frac{2}{4}$ d $\frac{2}{3}$ e $\frac{4}{5}$

2 Write the fractions shown.

a 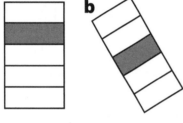 b c 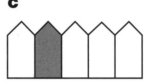 d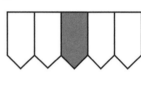

What do you notice?

_____

_____

3 Shade each rectangle to show $\frac{1}{4}$ in four different ways.

How do they all show $\frac{1}{4}$?

Explain your thinking.

_____

_____

4 Kaede says that the diagram below shows that $\frac{1}{4}$ is greater than $\frac{1}{2}$.

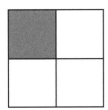 

Do you agree? ☐ Explain why.

_____

_____

5 Samson was told that these shapes showed quarters.

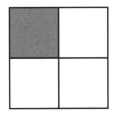   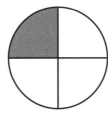   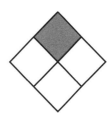

Shape A   Shape B   Shape C

He thinks they cannot be quarters because they all look different.

Do you agree? ☐ Explain why.

_____

_____

6 Nicole has made a picture. She says that the shaded part is one half of the whole.

Do you agree? ☐ Explain why.

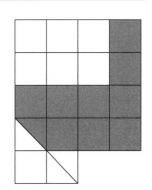

_____

_____

_____

Date: _____

Number

Lesson 4: **Fractions of quantities and sets**

- Understand that fractions describe equal parts of a quantity or set of objects

1 Draw a ring around half of each group of marbles.

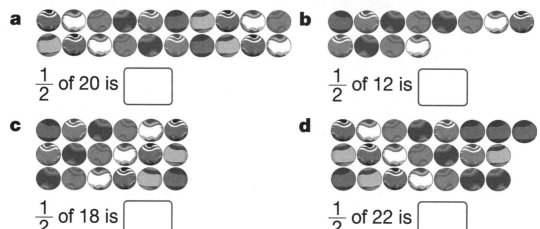

a

$\frac{1}{2}$ of 20 is ☐

b

$\frac{1}{2}$ of 12 is ☐

c

$\frac{1}{2}$ of 18 is ☐

d

$\frac{1}{2}$ of 22 is ☐

2 Ade cut some fruit into equal parts for his family to eat. Complete the table to show how much they have eaten.

Fruit	Number of equal parts altogether	Number of parts eaten	Fraction eaten
apple	4	2	$\frac{2}{4}$
melon	4	3	
pineapple	5	2	
orange	6	5	
pear			$\frac{2}{3}$
banana			$\frac{3}{5}$
watermelon			$\frac{4}{10}$

3 Freddie gives $\frac{1}{5}$ of his money to charity. He has $80 left.

How much money did he start with? Draw a part–whole model.

  4 In Stage 3, $\frac{1}{4}$ of the children are boys.

There are 30 girls.

How many children are there in Stage 3 altogether? Draw a part–whole model.

5 Use these digits to write six different fractions.

  2   3   4   5   6   10

Example: 2 parts out of a total of 3 parts = $\frac{2}{3}$

a [ ] parts out of a total of [ ] parts = [ ]

b [ ] parts out of a total of [ ] parts = [ ]

c [ ] parts out of a total of [ ] parts = [ ]

d [ ] parts out of a total of [ ] parts = [ ]

e [ ] parts out of a total of [ ] parts = [ ]

f [ ] parts out of a total of [ ] parts = [ ]

Date: _____   ☺ 😐 ☹

Lesson 1: **Same value, different appearance**

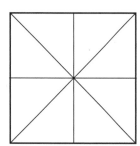

- Recognise that two fractions can have the same value

1 $\frac{2}{4}$ is the same as $\frac{1}{2}$. Colour $\frac{1}{2}$ of each shape.

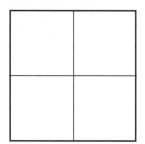 = 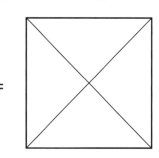 =

2 $\frac{4}{8}$ is also the same as $\frac{1}{2}$. Colour $\frac{4}{8}$ of each shape.

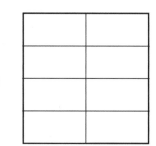 =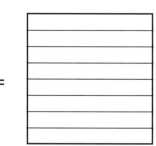

3 Colour each circle to show $\frac{1}{2}$. Then write the equivalent fraction.

a 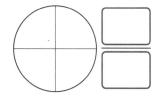 **b** 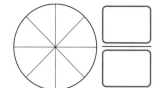 **c**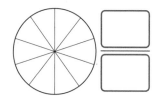

4 Colour each circle to show $\frac{1}{4}$. Then write the equivalent fraction.

a 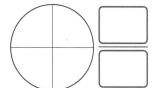 **b**  **c**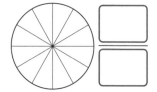

Number

5 Four friends each buy a pizza of the same size.

Ali cuts her pizza into 2 equal slices and eats 1 of them.

Ben cuts his pizza into 10 equal slices.

Caila cuts her pizza into 4 equal slices.

Deepak cuts his pizza into 8 equal slices.

How many slices must each person eat so that they eat the same as Ali?

a Ben must eat [ ] slices.

b Caila must eat [ ] slices.

c Deepak must eat [ ] slices.

6 Draw a ring around the shapes that do **not** have $\frac{1}{5}$ shaded.

a

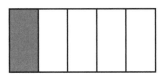

b

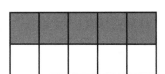

c

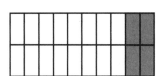

d

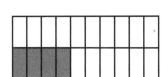

e

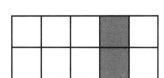

f

7 Draw a ring around the fractions that are equal to $\frac{1}{2}$.

②

$\frac{3}{6}$ $\frac{2}{4}$ $\frac{50}{100}$ $\frac{6}{10}$ $\frac{4}{8}$ $\frac{5}{7}$ $\frac{6}{12}$ $\frac{8}{9}$ $\frac{10}{20}$

What do you notice about the numbers in fractions that equal $\frac{1}{2}$?

_____

_____

Date: _____

Lesson 2: **Equivalent fractions**

Number

> • Recognise that two fractions can have the same value

1 Draw lines to match the equivalent fractions.

$\dfrac{1}{10}$ $\dfrac{2}{10}$

$\dfrac{1}{4}$ $\dfrac{2}{4}$

$\dfrac{1}{5}$ $\dfrac{2}{20}$

$\dfrac{1}{2}$ $\dfrac{2}{8}$

Explain how you know you are correct.

_____

_____

2 Draw a ring around the shapes that are shaded to show $\dfrac{1}{5}$.

a 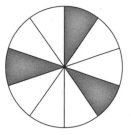   **b**    **c**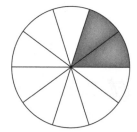

d 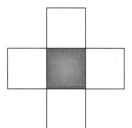   **e** 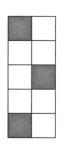   **f**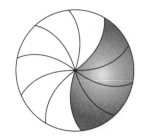

Number

 Nora thinks that $\frac{1}{2}$ and $\frac{1}{4}$ are equivalent fractions.

Do you agree? ▢

Explain why.

_____

 Vivaan says she has shaded $\frac{3}{4}$ of this shape.

Do you agree? ▢

Explain why.

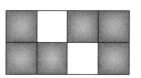

_____

 Complete these statements.

a $\dfrac{1}{4} = \dfrac{\boxed{}}{8} = \dfrac{\boxed{}}{12}$

b $\dfrac{1}{5} = \dfrac{\boxed{}}{10} = \dfrac{\boxed{}}{15}$

6 Complete the statement.

$\dfrac{\boxed{}}{\boxed{}} = \dfrac{2}{20} = \dfrac{\boxed{}}{\boxed{}} = \dfrac{\boxed{}}{\boxed{}} = \dfrac{5}{50}$

How did you work out the three missing fractions?

_____

_____

7 Tanya says: ◁ To find an equivalent fraction, you can just double the numerator and the denominator.

Is Tanya's statement always true, sometimes true or never true?

▢ Explain why.

_____

_____

Date: _____

137

Number

Lesson 3: **Comparing fractions**

• Compare two fractions

1 Draw a ring around the largest fraction in each pair.

a $\frac{1}{4}$ 　　b $\frac{1}{10}$ 　　c $\frac{1}{4}$

$\frac{1}{2}$ 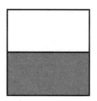　　$\frac{1}{8}$ 　　$\frac{1}{5}$

2 Write >, < or = between each pair of fractions.

a $\frac{3}{4}$ ☐ $\frac{1}{4}$　　　　b $\frac{3}{5}$ ☐ $\frac{4}{5}$

c $\frac{1}{5}$ ☐ $\frac{1}{2}$　　　　d $\frac{1}{6}$ ☐ $\frac{1}{9}$

e $\frac{1}{4}$ ☐ $\frac{2}{8}$　　　　f $\frac{6}{10}$ ☐ $\frac{3}{10}$

g $\frac{1}{10}$ ☐ $\frac{1}{8}$　　　　h $\frac{4}{5}$ ☐ $\frac{8}{10}$

i $\frac{2}{3}$ ☐ $\frac{3}{6}$　　　　j $\frac{1}{2}$ ☐ $\frac{5}{8}$

3 Barnie says: 2 is smaller than 5, so $\frac{1}{2}$ must be smaller than $\frac{1}{5}$.

Do you agree? ☐

Why?

_____

_____

Number

 4 Arjun thinks that $\frac{1}{2}$ is greater than $\frac{3}{4}$, because a half is bigger than a quarter.

Do you agree? ☐ Why?

_____

_____

5 Write a fraction to make the statement true.

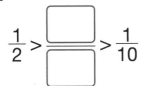

$\frac{1}{2} > \dfrac{\square}{\square} > \dfrac{1}{10}$

How many different fractions can you find?

┌───┐
│   │
│   │
│   │
│   │
└───┘

6 What fraction of the bar does each section represent? ☐

Draw two more bars of the same size and divide one into eighths and the other into sixths.

┌───┐
│   │
│   │
│   │
│   │
│   │
└───┘

Which fraction is the largest: a tenth, an eighth or a sixth?

┌──────────────────────┐
│   │
└──────────────────────┘

How do the bars help you to explain your answer?

_____

_____

Date: _____

139

Lesson 4: **Ordering fractions**

• Order a set of fractions

1 Order these fractions. Start with the smallest.

a $\frac{1}{2}, \frac{1}{4}, \frac{1}{10}$

b $\frac{4}{5}, \frac{1}{5}, \frac{3}{5}$

2 Write these fractions on the number line.

$\frac{3}{4}$ $\frac{1}{4}$ $\frac{2}{4}$

0 1

3 Order these fractions. Start with the smallest.

$\frac{7}{10}$ $\frac{2}{10}$ $\frac{5}{10}$ $\frac{3}{10}$ $\frac{6}{10}$ $\frac{1}{10}$

4 Colour the fraction bars so that the four fractions are in order, largest to smallest. Then write the fractions.

a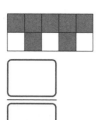

b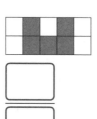

c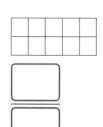

d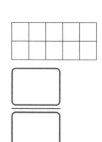

Number

5 Brenan says: To order fractions, you need to look at the denominator not the numerator.

Is this always, sometimes or never the way to order fractions?

[ ]   Why?

_____

_____

6 Ruby says: When I order fractions, I always order them according to the numerator.

Is this always, sometimes or never the way to order fractions?

[ ]   Why?

_____

_____

7 Anjani ordered some fractions, from largest to smallest.

This is the order she wrote: $\frac{1}{8}$ $\frac{1}{6}$ $\frac{1}{3}$ $\frac{1}{2}$

What is her mistake?

_____

_____

8 Write four fractions in the boxes on the number line.

[ ]
[ ]

0 |——————————————————————————————| 1

[ ] [ ]   [ ]
[ ] [ ]   [ ]

Date: _____

Lesson 1: **Fractions and division (A)**

• Understand that a fraction can be represented as a division

1 What fraction does everybody get?

a 1 dumpling shared between 2 people.

b 2 dumplings shared between 2 people.

c 3 dumplings shared between 2 people.

d 4 dumplings shared between 2 people.

2 There are 12 paper clips. You take 6 of them.

What fraction of the paper clips did you take?

How do you know? It might help to draw a diagram.

3 Mya has 3 pizzas. She wants to share them equally between herself and 4 friends.

What fraction of the pizzas will they each have?

How do you know? It might help to draw a diagram.

4 Sam has 15 apples. He gives 5 of them to a friend.

What fraction is left? $\dfrac{\Box}{\Box}$ How do you know? It might help to draw a diagram.

5 Aapti is given 16 note pads. She gives 4 of them to a friend.

What fraction does she have left? $\dfrac{\Box}{\Box}$ How do you know?
It might help to draw a diagram.

6 Jamie thinks that if he shares the $3 he has between 4 people, they will each get 75c.

Do you agree? $\Box$ Explain why. It might help to draw a diagram.

Date: _____   ☺ 😐 ☹

Number

Lesson 2: **Fractions and division (B)**

> • Recognise fractions as operators

1 Draw lines to match each fraction with its division calculation.

$\frac{1}{4}$ of 12	$12 \div 3$
$\frac{1}{3}$ of 30	$12 \div 2$
$\frac{1}{2}$ of 80	$30 \div 3$
$\frac{1}{2}$ of 36	$20 \div 4$
$\frac{1}{3}$ of 12	$80 \div 10$
$\frac{1}{10}$ of 20	$12 \div 4$
$\frac{1}{10}$ of 80	$36 \div 2$
$\frac{1}{4}$ of 20	$80 \div 2$
$\frac{1}{2}$ of 12	$20 \div 10$

2 Myesha has different piles of rice grains. She needs to find fractions of these amounts. Fill in the table to show the division calculations she should use to work out the answers.

Number of rice piles	Fraction needed	Division	Answer
20	$\frac{1}{2}$	$20 \div 2 =$	10
16	$\frac{1}{2}$		
24	$\frac{1}{4}$		
12	$\frac{1}{3}$		
30	$\frac{1}{10}$		
18	$\frac{1}{3}$		
12	$\frac{1}{4}$		
26	$\frac{1}{2}$		

Number

3 Write the fraction statement that matches each division calculation. You do **not** need to work out the answers.

Example: $32 \div 2 =$ $\boxed{\dfrac{1}{2} \text{ of } 32}$

a $20 \div 4 =$

b $14 \div 2 =$

c $15 \div 5 =$

d $24 \div 4 =$

e $30 \div 3 =$

f $32 \div 2 =$

g $25 \div 5 =$

h $18 \div 3 =$

4 Shay says:

> To find $\dfrac{1}{3}$ of 21, you divide 21 by 1 and multiply by 3.

Do you agree? ☐ Why?

_____

_____

5 Use the fraction and number cards to make questions. Then write each of them as a division and work out the answer.

Example: $\dfrac{1}{3}$ of $12 = 12 \div 3 = 4$

| 24 | $\dfrac{1}{10}$ of | 40 | $\dfrac{1}{2}$ of | 20 | $\dfrac{1}{3}$ of | 12 | $\dfrac{1}{4}$ of |

Date: _____

Lesson 3: **Adding and subtracting fractions (A)**

• Add and subtract fractions with the same denominator

1 Add these fractions.

a $\frac{1}{3} + \frac{1}{3} = \frac{\boxed{}}{\boxed{}}$

b $\frac{1}{5} + \frac{2}{5} = \frac{\boxed{}}{\boxed{}}$

c $\frac{1}{4} + \frac{3}{4} = \frac{\boxed{}}{\boxed{}}$

d $\frac{1}{8} + \frac{2}{8} = \frac{\boxed{}}{\boxed{}}$

e $\frac{1}{5} + \frac{3}{5} = \frac{\boxed{}}{\boxed{}}$

f $\frac{1}{8} + \frac{4}{8} = \frac{\boxed{}}{\boxed{}}$

2 Subtract these fractions.

a $\frac{3}{4} - \frac{1}{4} = \frac{\boxed{}}{\boxed{}}$

b $\frac{3}{4} - \frac{2}{4} = \frac{\boxed{}}{\boxed{}}$

c $\frac{7}{8} - \frac{1}{8} = \frac{\boxed{}}{\boxed{}}$

d $\frac{2}{3} - \frac{1}{3} = \frac{\boxed{}}{\boxed{}}$

e $\frac{7}{8} - \frac{3}{8} = \frac{\boxed{}}{\boxed{}}$

f $\frac{3}{5} - \frac{2}{5} = \frac{\boxed{}}{\boxed{}}$

3 Anya says: I can add $\frac{1}{4}$ and $\frac{1}{4}$. My answer is $\frac{2}{8}$.

Do you agree? ☐ Explain your thinking.

_____

_____

4 Pedro says: I can subtract fractions with the same denominators. All I need to do is find the difference between the numerators.

Do you agree? ☐ Why?

_____

_____

Number

5 Ella eats $\frac{4}{8}$ of a pizza.

Tom eats $\frac{1}{8}$ of it. What fraction do they eat altogether?

What fraction is left? Explain your thinking. You can draw a diagram.

6 Suzie folds a strip of paper into fifths.

She tears off 3 parts. What fraction is left?

Explain your thinking. You can draw a diagram.

7 Sam eats $\frac{1}{4}$ of a pizza. Adnam eats $\frac{1}{2}$ of the pizza. Zac eats the rest of the pizza.

What fraction of the pizza does Zac eat?

Draw a diagram to show how much pizza Sam, Adnam and Zac each eat.

How does your diagram help you to explain your reasoning?

_____

_____

Date: _____

Number

Lesson 4: **Adding and subtracting fractions (B)**

• Add and subtract fractions with the same denominator

1 Add these fractions.

a $\dfrac{1}{6} + \dfrac{1}{6} = \dfrac{\Box}{\Box}$

b $\dfrac{1}{7} + \dfrac{2}{7} = \dfrac{\Box}{\Box}$

c $\dfrac{1}{9} + \dfrac{3}{9} = \dfrac{\Box}{\Box}$

d $\dfrac{1}{10} + \dfrac{2}{10} = \dfrac{\Box}{\Box}$

e $\dfrac{2}{7} + \dfrac{3}{7} = \dfrac{\Box}{\Box}$

f $\dfrac{3}{9} + \dfrac{4}{9} = \dfrac{\Box}{\Box}$

2 Subtract these fractions.

a $\dfrac{3}{6} - \dfrac{1}{6} = \dfrac{\Box}{\Box}$

b $\dfrac{3}{7} - \dfrac{2}{7} = \dfrac{\Box}{\Box}$

c $\dfrac{7}{9} - \dfrac{2}{9} = \dfrac{\Box}{\Box}$

d $\dfrac{7}{10} - \dfrac{4}{10} = \dfrac{\Box}{\Box}$

e $\dfrac{5}{7} - \dfrac{3}{7} = \dfrac{\Box}{\Box}$

f $\dfrac{3}{10} - \dfrac{2}{10} = \dfrac{\Box}{\Box}$

3 Kalpana says:

> If I cut my cake into six parts and eat two of them, I will have $\dfrac{4}{6}$ left.

Do you agree with Kalpana? ☐

Explain your thinking. You can draw a diagram.

4 Bourey says:

> I cut a rope into ten pieces. I give half of the pieces to my friend. I am left with $\dfrac{5}{10}$.

Do you agree? ☐

Explain your thinking. You can draw a diagram.

5 Carlos eats $\frac{7}{10}$ of a pie. Tom eats $\frac{1}{10}$ of it. What fraction do they

eat altogether? ⬚ What fraction is left? ⬚

Explain your thinking. You can draw a diagram.

6 Jo has a length of material. She cuts off $\frac{1}{5}$. Then she cuts off another tenth. What is the total fraction that she has cut off?

⬚ What fraction is left? ⬚ Explain your thinking.

You can draw a diagram.

7 Lily eats $\frac{3}{10}$ of a chocolate bar. Florence eats $\frac{1}{2}$ of the chocolate bar. Sue eats the rest of the chocolate bar.

What fraction of the chocolate bar does Sue eat? ⬚

Draw a diagram to show how much chocolate Lily, Florence and Sue each eat.

Date: _____

☺ 😐 ☹

Geometry and Measure

Lesson 1: **Units of time**

• Choose suitable units to measure time

1 **a** Draw lines to match each picture to its name label.

stopwatch   analogue clock   digital clock

b Write one place you might see each one being used.

stopwatch: _____

analogue clock: _____

digital clock: _____

2 What can you do in about one minute? Write three different things.

_____

_____

_____

3 Bertie thinks that it will take him one minute to read the 12 pages he has left in his book.

Is this a good estimate? ☐ Why?

_____

_____

4 Elsa thinks it will take her 3 hours to eat her breakfast.

Is this a good estimate? ☐ Why?

_____

_____

Geometry and Measure

5 Bobby is going to spend a week with his cousin.
He thinks that he will be with his cousin for seven days.

Do you agree? ☐ Why?

_____

_____

6 Match the suitable unit of time.

sunlight in a day   seconds

clap your hands   years

grow a tree   minutes

listen to a song   months

spring   hours

7 Work out the time intervals and match the units of time.

Thursday 2nd to Tuesday 7th   ☐ weeks

1st March, 2021 to 1st July, 2021   ☐ years

7th February to 21st February   ☐ months

2019 to 2022   ☐ days

8 Look at this page from a calendar.

a How many weeks from 8th to 22nd? ☐

b How many days is that? ☐

c How many days from 3rd to 31st? ☐

d How many weeks is that? ☐

JANUARY   2020

Mon	Tue	Wed	Thu	Fri	Sat	Sun
		1	2	3	4	5
6	7	8	9	10	11	12
13	14	15	16	17	18	19
20	21	22	23	24	25	26
27	28	29	30	31		

Date: _____

Geometry and Measure

Lesson 2: **Telling the time (A)**

* Read and write the time to five minutes

1 Draw the times on the clock faces.

a 6 o'clock **b** half past 10 **c** 20 minutes past 9

d 45 minutes past 2 **e** 35 minutes past 3 **f** $\frac{1}{4}$ past 11

2 Write the digital time to match each analogue clock.

a [:] **b** [:] **c** [:]

3 Write the digital time to match each analogue clock.

a [:] **b** [:] **c** [:] **d** [:]

Draw the times on the analogue clocks.

  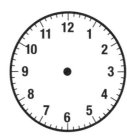 

a 2:30   **b** 10:45   **c** 8:05   **d** 12:40

5  Chenglei says:

5:10 is the same time as 5 minutes past 10.

Do you agree? ☐

Explain why.

_____

6 The analogue clock shows 50 minutes past 5. The digital clock shows 5:50. The clocks show the same time. We can also read this time as 10 minutes to 6.

What are the 3 ways that we can read these times?

a    **b**    **c**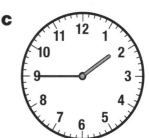

_____   _____   _____

_____   _____   _____

_____   _____   _____

Geometry and Measure

Date: _____

Lesson 3: **Telling the time (B)**

Geometry and Measure

> • Read and write the time

1 Draw the times on the clock faces.

a 17 minutes past 3   **b** 32 minutes past 11   **c** 48 minutes past 6

d 3 minutes past 10   **e** 56 minutes past 2   **f** 21 minutes past 5

2 Write the digital time to match each analogue clock.

a [:]   **b** [:]   **c** [:]   **d** [:]

3 Draw the times on the analogue clocks.

a 3:16   **b** 10:24   **c** 4:39   **d** 8:07

4

Sasha says: The clock shows 10 minutes to 3.

Ben says: The hour hand is not quite pointing to the 3, so it must be 10 minutes to 2.

Who is right? [ ] Why?

_____

_____

5 a Draw lines to match the clocks that show the same time.

| 3:46 | 2:36 | 9:16 | 7:12 |

b Does it matter that there are no hour numbers? [ ] Explain your thinking.

_____

_____

6 This clock has lost its minute hand.

What time could it be? [ ]

Is there another possibility? [ ]

Explain why.

_____

_____

Date: _____

155

Lesson 4: **Timetables**

- Use and interpret timetables
- Find time intervals

Geometry and Measure

1 Look at the timetable and answer the questions.

9:00 – 10:00 Maths	11:00 – 12:00 English
10:00 – 10:30 Reading	12:00 – 1:00 Lunch
10:30 – 11:00 Break	1:00 – 3:00 Science

a What time does English begin?

b What lesson is in the afternoon?

c What time is break?

d How long is lunch?

e What time does Maths finish?

f What time does Science finish?

2 Look at the timetable and answer the question. All the times are in the morning.

	Leaves at	Arrives at
Coach A	7:30	8.00
Coach B	7:45	8.45
Coach C	8:00	9.00
Coach D	9.00	9.45

What time did each coach arrive?

Coach A	
Coach B	
Coach C	
Coach D	

Geometry and Measure

3 Carlos wants to catch the 2:00 bus to town. The bus journey takes 1 hour. Carlos thinks he will get to town at 2:30.

Do you agree? ☐

Explain. _____

_____

4 Macy spent 2 hours and 30 minutes reading her book.
George spent 2 hours and 45 minutes reading his book.
Ruby spent 2 hours and 20 minutes reading hers.

Who read for the longest time?_____

Who read for the shortest time?_____

5 The timetable shows the journey times of four trains leaving Station A and arriving at Station B.

	Station A	Station B
Train 1	10:55	11:55
Train 2	11:30	1:30
Train 3	12:40	2:40
Train 4	1:55	3:55

a Write the journey time of each train.

Train 1 ☐   Train 2 ☐

Train 3 ☐   Train 4 ☐

b Which train has the longest journey time? ☐

c Which train has the shortest journey time? ☐

Date: _____

Geometry and Measure

Lesson 1: **2D shapes**

• Identify, describe, name and sketch regular and irregular 2D shapes

You will need
• coloured pencils

1 Label the shapes. Choose the names from the box.

triangle	irregular pentagon	regular pentagon
rectangle	regular hexagon	irregular hexagon
	regular octagon	semi-circle

_____ _____ _____ _____

_____ _____ _____ _____

2 What 2D shapes can you see in this picture?

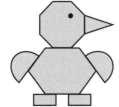

_____

_____

_____

_____

3 a Draw a shape picture using different shapes. Colour each shape a different colour.

b Write the names of the shapes you used.

_____

_____

_____

Geometry and Measure

4 Look at the shapes.

a Draw a ring around a **regular** shape and describe its properties.

_____

b Draw a cross on an **irregular** shape and describe its properties.

_____

5 Which of these shapes is the odd one out?

Why? _____

_____

Think of another shape that could be the odd one out. Explain why.

_____

_____

6 a Draw a line to split this octagon into a right-angled hexagon and a 4-sided shape.

b Put a cross next to the right angles you have made in your hexagon.

c How do you know it is a hexagon?

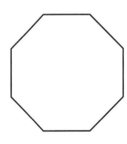

_____

7 Write a definition of a vertex.

_____

_____

Date: _____

Lesson 2: **Sorting 2D shapes**

- Sort regular and irregular 2D shapes

You will need
- ruler

1 Draw lines to complete these shapes.

pentagon   semi circle   octagon   hexagon   triangle

2 Samira sorted some shapes and recorded them in this table.

Write in the headings to show how she sorted them.

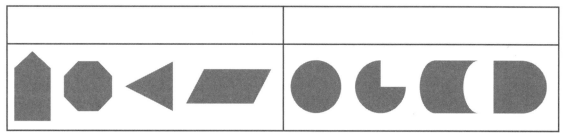

3 Draw a semi circle and describe its properties. Include these words:

curved   straight   side   non-polygon

_____

_____

_____

_____

4 a Name these regular shapes.

_____ _____ _____ _____

b Draw an irregular shape for each of the shapes in part **a**.

Geometry and Measure

5 Xiang is thinking of some shapes and recording them in a Carroll diagram.

Draw some of the shapes he could have in his Carroll diagram.

Draw four possible shapes in each section.

Regular shapes	Not regular shapes

6 Draw a hexagon with two right angles.

7 Choose your own way to sort these shapes into a table. Give each section of the table a heading. Copy the shapes into the correct section of the table.

Date: _____

Lesson 3: **Symmetry**

- Identify horizontal and vertical lines of symmetry on 2D shapes

You will need
- ruler

1 Draw one line of symmetry on each of these shapes.

2 This is half of a symmetrical shape. Draw the other half to make the whole shape.

3 Draw two lines of symmetry on each of these shapes.

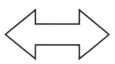 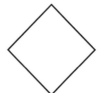 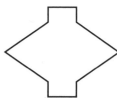

4 What is symmetry?

_____

_____

_____

5 Draw all the lines of symmetry on these shapes.

6 Circle the shape that is the odd one out.

Explain why it is the odd one out. _____

_____

Geometry and Measure

7 Complete this pattern so that it is symmetrical.

line of symmetry

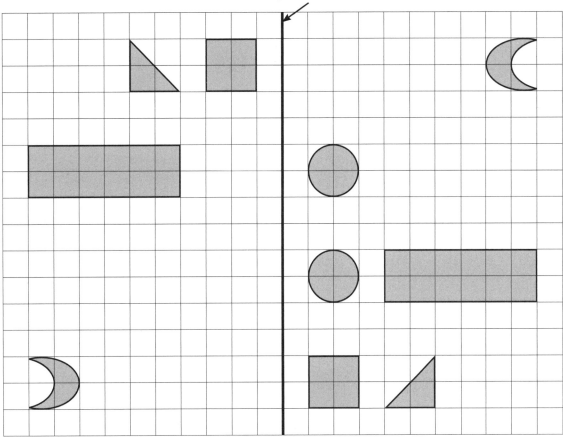

8 Now make your own symmetrical pattern on this grid.

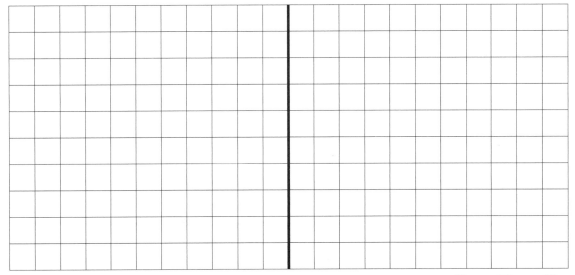

Date: _____

163

Lesson 4: **Angles**

- Compare angles with a right angle
- Recognise that a straight line is equal to two right angles or a half turn

You will need
- ruler

1 Circle the right angles.

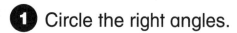

2 Draw a ring around the right angles in these pictures.

3 a What size of turn is a right angle?

_____

b What size of turn is 2 right angles?

_____

4 Circle the angle that is the odd one out.

Explain why it is the odd one out.

_____

5 Draw a ring around the angles **greater** than a right angle.

6 Use this shape as the centre shape for a pattern of tiles.
Make sure some of your tile shapes have right angles.

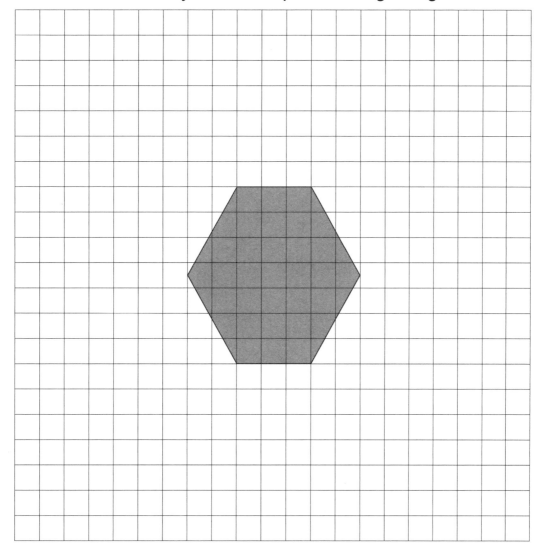

7 What other 2D shapes did you use in your pattern in **6**?

_____

_____

Date: _____

Geometry and Measure

Lesson 1: **Identifying 3D shapes**

- Identify, describe, classify, name and sketch 3D shapes
- Recognise pictures, drawings and diagrams of 3D shapes

1 Draw lines to match each 3D shape with its name.

sphere   square-based pyramid   cylinder   cuboid   cube

   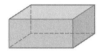

2 What are the properties of a square-based pyramid?

_____

_____

3 Write the names of the shapes on the Carroll diagram.

6

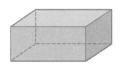

	square faces	no square faces
12 edges		
not 12 edges		

Geometry and Measure

4 Circle the shape that is the odd one out.

  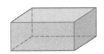

Explain why it is the odd one out.

_____

_____

Draw another shape that is an odd one out.

Explain why it is an odd one out.

_____

_____

_____

5 Katie is thinking of a shape. She says:

My shape has 4 faces, 6 edges and 4 vertices.

What is Katie's shape? [   ]

6 Marley has drawn a square. He says it is the face of a 3D shape. What shape could he be thinking of? Think of three possibilities. Sketch them and write the name of each 3D shape.

Date: _____

Lesson 2: **Prisms**

- Identify, describe, classify, name and sketch prisms
- Recognise pictures, drawings and diagrams of prisms

1 Draw a ring around all the prisms, then write the name of each shape.

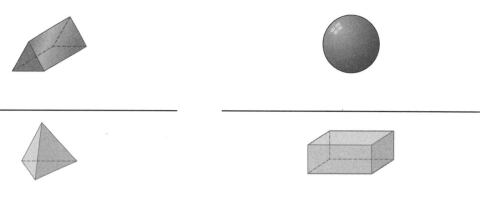

_____   _____

_____   _____

2 Jayesh sorted five 3D shapes and recorded them in this table.

Write in the headings to show how he sorted them.

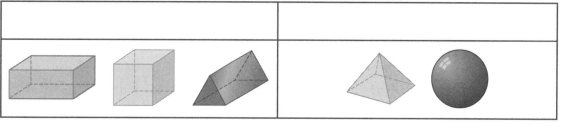	

3 Draw lines to match each 3D shape to its name.

sphere   cuboid   cube   cylinder   triangular prism

Geometry and Measure

4 **a** Draw a ring around all the cubes.

  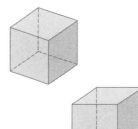

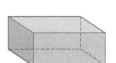 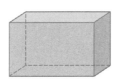

b Complete the table to show the properties of a cube.

Number of faces	
Number of vertices	
Number of edges	
Is it a prism?	

c Describe how a cube is different from a cuboid.

_____

_____

5 Look closely at a cuboid.
Then draw and label it.

6 Look at these shapes.
What is the same about them?
Write three similarities.

_____

_____

_____

What is different? Write three differences.

_____

_____

_____

Date: _____

Lesson 3: **Pyramids**

- Identify, describe, classify, name and sketch pyramids
- Recognise pictures, drawings and diagrams of pyramids

1 Draw a ring around the pyramids.

2 Explain how you know they are pyramids.

_____

3 Ben thinks a pyramid is made from triangular faces.

Is this always, sometimes or never true? _____

Explain why.

_____

_____

4 'Triangular pyramids and triangular prisms have the same number of faces.'

Is this statement true or false? _____

Explain why.

_____

_____

5 Khadija says:

Do you agree? _____

Explain why.

> I think that a hexagonal-based pyramid has 7 faces.

_____

_____

6 Ranjit is describing a 3D shape.

He says that some of its faces are triangles.

What shape could Ranjit be describing?

Think of three possibilities.

Sketch them and write their names.

Date: _____

Lesson 4: **3D shapes in real life**

- Identify, describe, classify, name and sketch 3D shapes
- Recognise pictures, drawings and diagrams of 3D shapes

1 Name a real-life object that is the shape of a cube.

_____

2 Name a real-life object that is the shape of a sphere.

_____

3 Where might you see these shapes in real life?

cuboid   cylinder

_____   _____

_____   _____

_____   _____

4 Write the names of the 3D shapes you can see in the picture.

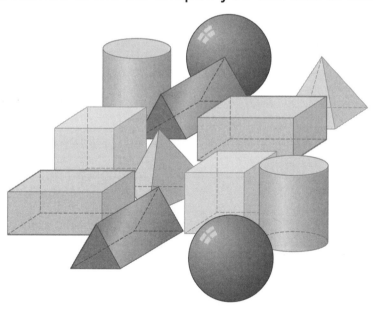

_____

_____

_____

Geometry and Measure

172

5 Make a list of 3D objects that you can see in the classroom.
Then write the 3D shape name for each object.

Object	3D shape name

6 Draw a ring around the object that is the odd one out.

Explain why.

_____

7 **a** Use 3D shapes to build a real-life
model of this shape.

b What 3D shapes did you use?

_____

_____

Could you have used any others?

_____

_____

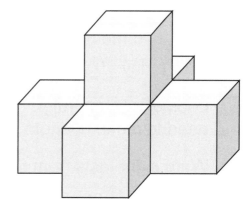

Date: _____

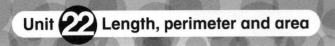

Lesson 1: **Units of length**

Geometry and Measure

- Estimate and measure lengths in centimetres (cm), metres (m) and kilometres (km)
- Understand the relationship between units

You will need
- metre stick
- ruler

1 How many centimetres are there in 4 m? ☐ cm

Show how you worked out your answer.

☐

2 How many metres are there in 2 km? ☐ m

Show how you worked out your answer.

☐

3 Order these lengths, from shortest to longest.

10 km   5 cm   12 m   3 km   25 cm   41 m

4 Look at a metre stick.

How many 10 cm lengths are there in 1 m? ☐

5 Look at a 30 cm ruler. How many would you need to make a length a little longer than 1 m? ☐

6 Write eight facts from 5 m = 500 cm.

Example: 10 m = 1000 cm

☐ ☐ ☐ ☐

☐ ☐ ☐ ☐

Explain the strategies you used to work these out.

_____

_____

_____

7 Write six facts from 1 km = 1000 m.

Explain the strategies you used to work these out.

_____

_____

8 **a** Dana wondered what length she would have if she converted 3 km and 3 m into metres. What do you think?
Show your answer and how you worked it out. [ ]

[   ]

b How many centimetres would that be? [ ]

9 Write eight facts from $4\frac{1}{2}$ m = 450 cm. Be creative!

Explain the strategies you used to work these out.

_____

_____

_____

Date: _____   ☺ ☺ ☹

Lesson 2: **Measuring lines**

Geometry and Measure

• Estimate and measure lengths in centimetres (cm)
• Use instruments that measure length

You will need
• ruler
• paper
• scissors

1 Estimate the lengths of these lines. Write each estimate on the line. Then use your ruler to measure the lines. Write each measurement in centimetres.

[ ] cm

[ ] cm

[ ] cm

[ ] cm

[ ] cm

2 Use your ruler to draw a line 11 cm long.

3 Use your ruler to draw a line 8 cm long.

4 Saul draws a line. It is 6 cm long.

   a Use your ruler to draw a line half the length of Saul's line.

   b Use your ruler to draw a line twice the length of Saul's line.

   c Use your ruler to draw a line 9 cm longer than Saul's line.

Geometry and Measure

5 Estimate the lengths of these lines. Write each estimate on the line. Then measure the lines. Write the length of each one, rounded to the nearest centimetre.

⊢————————————⊣   [ ] cm

⊢———————————————————⊣   [ ] cm

⊢——————————————————————⊣   [ ] cm

⊢——————————————⊣   [ ] cm

⊢———⊣   [ ] cm

6 Sophia says: ◁ To measure a line, all you do is put the end of your ruler at the beginning of the line and then see how long it is.

Is this a good explanation of how to use a ruler? [ ] Why?

_____

_____

7 Draw six lines of different length on strips of paper. Use a straight edge, not a ruler, so you don't know how long your lines are.
Cut the strips of paper to the length of your lines. Label each strip from A (shortest) to F (longest).
Draw a table like the one below and label the first column A to F. Then estimate the length of each strip and write it in the table. Next, measure each strip and write it in the table.
Finally, work out the difference between your estimates and the actual lengths.

Strip	Estimate	Actual	Difference
A			
B			

Date: _____   ☺ ☻ ☹

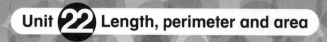

Lesson 3: **Perimeter**

- Understand that perimeter is the total distance around a 2D shape and can be calculated by adding lengths

You will need
- squared paper
- ruler

1 What is the perimeter of each shape?

a

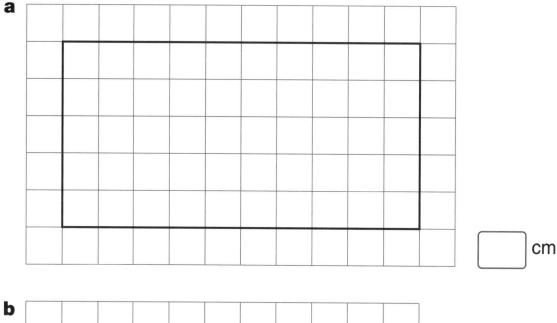

⬚ cm

b

⬚ cm

2 Explain how you worked out the perimeters in **1**.

_____

_____

3 Using a sheet of squared paper, draw each rectangle and work out its perimeter.

a 6 cm long and 3 cm wide

b 10 cm long and 1 cm wide

c 5 cm long and 5 cm wide

Geometry and Measure

4 Work out the perimeter of each shape.

Write the perimeter in the middle of the shape.

NOTE:
Shapes are not drawn to full scale.

12 cm

8 cm

5 cm

4 cm

5 cm

7 cm

6 cm

3 cm

5 **a** How many different rectangles with a perimeter of 20 cm can you draw?

Draw them on a piece of squared paper. Make sure you use whole squares.

b Work carefully to make sure you draw them all!

What pattern do you notice?

_____

_____

6 The length of a rectangle is double its width.

a If the perimeter is 12 cm, what are the length and width of

the rectangle? length: ☐ width: ☐

b What if the perimeter is 24 cm? length: ☐ width: ☐

c What if the perimeter is 30 cm? length: ☐ width: ☐

Use squared paper to work out your ideas. length:

Date: _____

Geometry and Measure

Lesson 4: **Area**

• Understand that area is how much space a 2D shape occupies within its perimeter

You will need
• ruler

1 Work out the area of each rectangle.

a
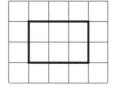

☐ square units

b
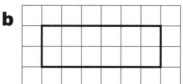

☐ square units

c
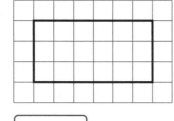

☐ square units

d
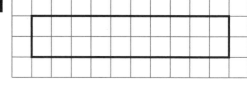

☐ square units

2 Explain how you worked out the areas.

_____

_____

3 Toby decides to make a patio.

He uses 16 square paving slabs.

He decides to lay 4 rows of 4 squares.

a Draw a diagram of Toby's patio.

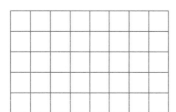

b What is the area of his patio? ☐ square units

Geometry and Measure

c Draw another design for Toby's patio.

d What is the area of your design?

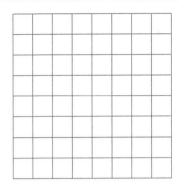

[ ] square units

4 Draw three different rectangles, each with an area of 12 square units. Label your shapes a, b, c.

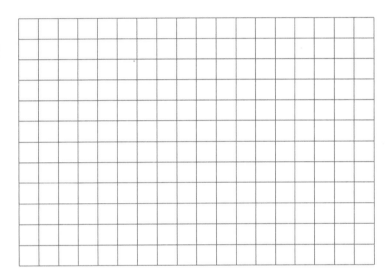

5 Look at your shapes from **4**. What are their perimeters?

a [ ]   **b** [ ]   **c** [ ]

6 What is the area of each shape?

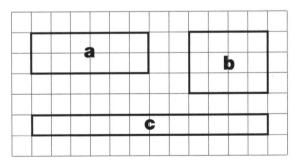

a [ ] square units

b [ ] square units

c [ ] square units

What does this tell us about area?

_____

_____

Date: _____

181

Lesson 1: **Units of mass**

Geometry and Measure

> - Estimate and measure mass in kilograms and grams
> - Understand the relationship between units

1 Write how many grams there are in:

a 1 kg

b 3 kg

c 7 kg

[   ] g   [   ] g   [   ] g

2 Write how many kilograms there are in:

a 2000 g [   ] kg   **b** 9000 g [   ] kg

c 4000 g [   ] kg   **d** 6000 g [   ] kg

3 a Order these masses, from heaviest to lightest.

$\frac{1}{2}$ kg   800 g   2 kg   100 g   50 kg   50 g

_____

b How many 100 g masses would balance 1 kg? [   ]

c How many 10 g masses would balance 1 kg? [   ]

4 Write the total mass for each set of weights, in grams.

a

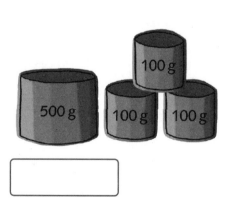

[   ]

b

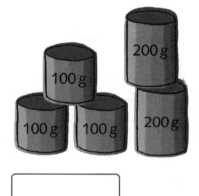

[   ]

Geometry and Measure

5 Draw lines to match the equivalent masses.

8 kg   5000 g   500 g $\frac{1}{4}$ kg

250 g   8000 g   5 kg $\frac{1}{2}$ kg

6 What mass will Ciara have if she converts 7 kg and 750 g, to grams? Show your working.

7 Make each mass, using some of these weights.

a 500 g

b 900 g

c 650 g

d 1 kg

8 Georgie says: All you need to do to change kilograms into grams is to add three zeros.

Is this a good explanation?   Why?

_____

_____

Date: _____

Lesson 2: **Measuring in kilograms**

Geometry and Measure

- Estimate and measure mass in kilograms
- Use instruments that measure mass

You will need
- 1 kg bag of rice
- kitchen scales

1 **a** Circle the things you think have a mass of **more** than 1 kg.

b Why do you think these things are measured in kilograms?

_____

2 Find six objects in the classroom you would measure in kilograms.

_____   _____

_____   _____

_____   _____

3 Estimate the mass of three objects in the classroom that have a mass of approximately 1 kg. Use a 1 kg bag of rice to help you. Write the name of each object and your estimate of its mass.

Object	Estimated mass

4 Now use a set of kitchen scales to find the masses of the three objects to the nearest $\frac{1}{2}$ kg.

Object	Mass to the nearest $\frac{1}{2}$ kg

Geometry and Measure

5 Read the mass on each scale.

a

b

c

6 Draw pointers on the scales to show these masses.

a 9 kg

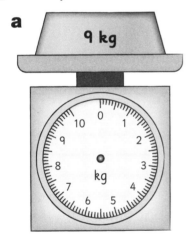

b 2 kg 500 g

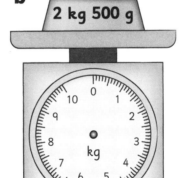

c 8 kg 500 g

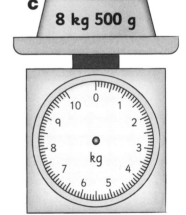

7 Write a sentence to describe what we mean by **mass**.

_____

_____

8 Kelly has one of each of these weights.
She uses a balance scale to find
the mass of a bag of apples.
She uses all the weights.

What is the mass of the apples? [ ]

Date: _____

Lesson 3: **Measuring in grams**

- Estimate and measure mass in grams
- Use instruments that measure mass

1 **a** Circle the things you think have a mass of **less** than 1 kg.

b What unit would you use to find the mass of the things you circled?

_____

2 Find six objects in the classroom you would measure in grams.

_____   _____

_____   _____

_____   _____

3 The weights on this scale total 700 g.

Using the weights to the right of the scale, what different weights could you use to make 700 g? Give three examples.

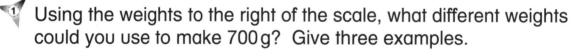

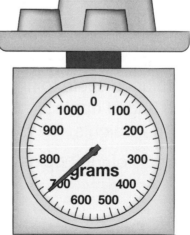

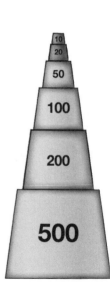

4 Read the mass on each scale.

a   b   c

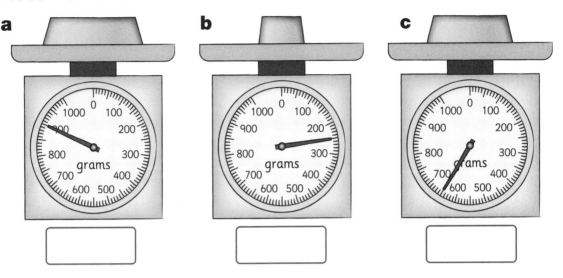

5 Draw pointers on the scales to show these masses.

a   400 g   b   950 g   c   350 g

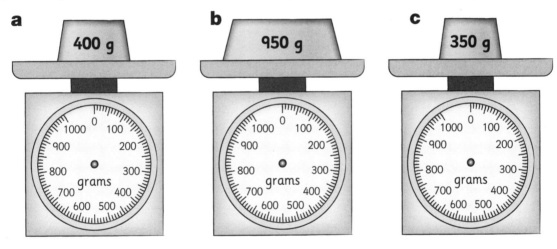

6 Samson measures a ball of clay. It is 200 g.

a What is the mass of a ball of clay three times the mass of Samson's?

b What is the mass of a ball of clay half the mass of Samson's?

c What is the mass of all three balls of clay?

Date: _____

Lesson 4: **Measuring in kilograms and grams**

- Estimate and measure mass in kilograms and grams
- Understand the relationship between units
- Use instruments that measure mass

1 Ali buys 500 g of beans and 1 kg of potatoes. Draw the pointer on the scale to show the total mass.

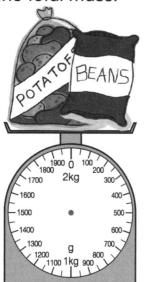

2 Dana buys 1 kg of rice, 250 g flour, 250 g lentils and 250 g sugar. Draw the pointer on the scale to show the total mass.

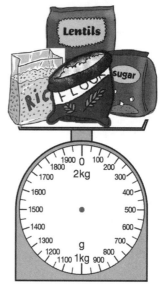

3 What is the total mass of these weights, in grams?

Show how you worked out your answer.

Geometry and Measure

Geometry and Measure

4 Sally says:

Do you agree?

If I know that 1000 g equals 1 kg, I can find many different equivalent masses. For example, 2400 g = 2 kg 400 g.

Write three examples.

5 Write four different ways to make 4 kg 250 g, using some or all of these weights.

6 Anita buys 1 kg 600 g of flour. Olivia buys twice as much.

How much does Olivia buy?

Show how you worked out your answer.

7 Yuko and Samuel buy a total of 2 kg 400 g of flour. Yuko buys $\frac{1}{4}$ and Samuel buys $\frac{3}{4}$. What mass do they each buy?

Yuko: Samuel:

Show how you worked out your answer.

Date: _____

Lesson 1: **Units of capacity**

- Estimate and measure capacity in litres and millilitres
- Understand the relationship between units

Geometry and Measure

1 Write how many millilitres there are in:

a 4 *l*   **b** 8 *l*   **c** 7 *l*   **d** 3 *l*

[ ] m*l*   [ ] m*l*   [ ] m*l*   [ ] m*l*

2 Write how many litres there are in:

a 6000 m*l* [ ] *l*   **b** 7000 m*l* [ ] *l*

c 5000 m*l* [ ] *l*   **d** 9000 m*l* [ ] *l*

3 a Order these capacities, from largest to smallest.

3 *l*   900 m*l* $\frac{1}{4}$ *l*   400 m*l* $\frac{1}{2}$ *l*   1 *l*

_____

b How many 100 m*l* is the same as 1 *l*? [ ] m*l*

c How many 10 m*l* is the same as 1 *l*? [ ] m*l*

4 Answer each question. Show your working.

a How many 200 m*l* jugs of water will it take to fill a 1 *l* bottle?

[ ]   [ ]

b How many 5 *l* buckets of water will it take to fill an 50 *l* bath?

[ ]   [ ]

c How many times would you need to fill a 1 *l* jug to empty a 10 *l* barrel of water?

[ ]   [ ]

5 Stefan says: Capacity is about water in containers.

Is this a good explanation? ☐

Why?

_____

_____

6 Draw a ring around the capacity that is the odd one out.

2500 m*l*   3600 m*l*   15 litres   5600 m*l*   5700 m*l*

Why is it the odd one out?

_____

7 Sienna fills three jugs with juice. The smallest jug holds 100 m*l*.
The middle-sized jug holds twice as much. The largest jug holds
4 times as much as the smallest.

What is the total amount of juice? ☐ Show your working.

8 Make as many different totals as you can, using two of
these amounts.

250 ml   500 ml   125 ml   1 *l*

Date: _____

Geometry and Measure

191

Lesson 2: **Measuring capacity**

- Estimate and measure capacity in litres and millilitres
- Understand the relationship between units

1 a Circle the items that you think have a capacity of **less** than 1 litre.

b What unit would you use to measure the capacity of the items you circled?

_____

2 a Circle the items that you think have a capacity of **more** than 1 litre.

b Why do you think these items are measured in litres?

_____

3 Sam has 1 litre of juice in a jug. He pours 500 m*l* into a bottle.

He pours 200 m*l* into a glass.

How much is left in the jug?

Explain how you worked this out.

_____

_____

4 Holly says:
The tallest container has the largest capacity.

 Is Holly's statement always true, sometimes true,
never true? [ ] Explain why.

_____

_____

5 Marie has 3 bottles of water with 500 ml in each.
Lee has one bottle of water with 1 and a half litres in it.

Who has the most water? _____

Explain why.

_____

_____

6 Calculate the capacity. Remember to use the correct unit of
measurement in your answer.

a 5 *l* = capacity of each  = [ ]

b 1 *l* = capacity of each 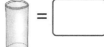 = [ ]

c 500 ml =  capacity of each 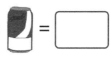 = [ ]

7 These measurements are in order, from least to most. Cross out
the measurements that are in the wrong place.

a 50 ml   100 ml   1 *l*   300 ml   750 ml   1 *l*   250 ml   3 *l*

b 7 *l*   12 *l*   25 *l*   30 ml   32 *l*   500 ml   38 *l*   100 *l*

Date: _____

Geometry and Measure

Geometry and Measure

Lesson 3: **Measuring in litres and millilitres**

- Estimate and measure capacity in litres and millilitres
- Understand the relationship between units
- Use instruments that measure capacity

1 How much liquid is in each jug?

You will need
- ruler

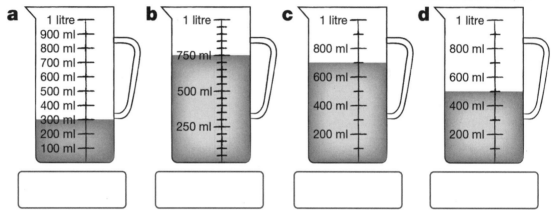

a   1 litre   900 ml   800 ml   700 ml   600 ml   500 ml   400 ml   300 ml   200 ml   100 ml

b   1 litre   750 ml   500 ml   250 ml

c   1 litre   800 ml   600 ml   400 ml   200 ml

d   1 litre   800 ml   600 ml   400 ml   200 ml

2 Shade the measuring cylinders to show:

   a   300 m*l*   **b**   1 *l* 500 m*l*   **c**   800 m*l*   **d**   1 *l* 700 m*l*

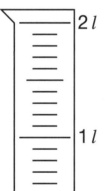

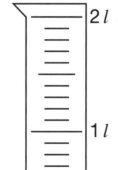

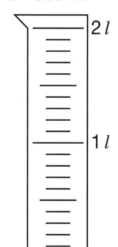

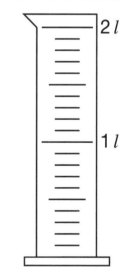

3 Daisy measures the capacity of a tub. She notices that the capacity is between 2 *l* 400 m*l* and 2 *l* 500 m*l*.

Give three possible capacities.

Geometry and Measure

4 The capacity of the watering can is 10 litres.

Jo uses 3*l* 500 m*l* to water her plants.

Show how much water is left in the watering can.

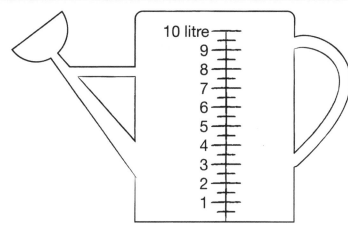

10 litre
9
8
7
6
5
4
3
2
1

5 Convert these measurements. Complete the table.

Millilitres	Litres and millilitres
5800 m*l*	
	7*l* 500 m*l*
8100 m*l*	
	4*l* 200 m*l*

6 Sharma is making a drink for a picnic. She uses 300 m*l* of apple cordial and 1200 m*l* of water. Draw a measuring cylinder. Draw the scale on the cylinder and show the total amount of juice that Sharma has made.

Date: _____

Lesson 4: **Temperature**

Geometry and Measure

• Use instruments that measure temperature

You will need
• coloured pencil
• paper

1 What temperatures are shown on these thermometers?

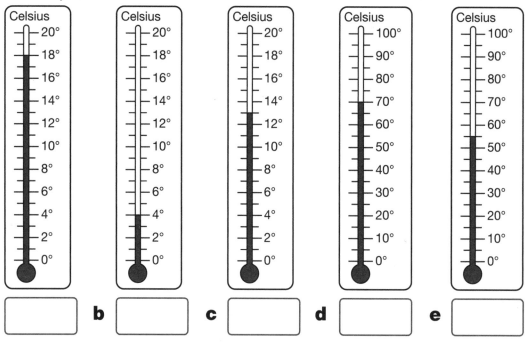

a [　　　] b [　　　] c [　　　] d [　　　] e [　　　]

2 Shade each thermometer to show the correct temperature.

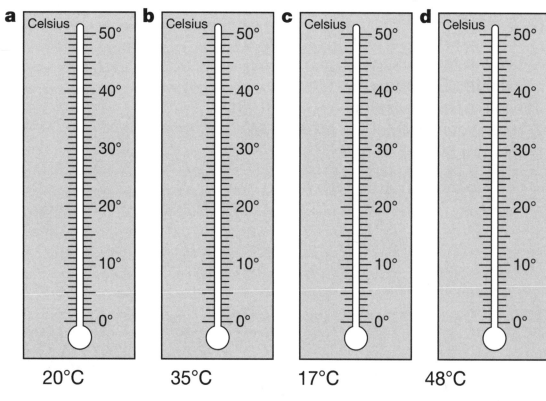

a 20°C   b 35°C   c 17°C   d 48°C

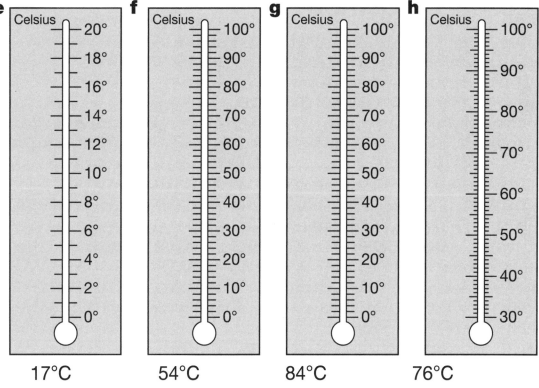

e 17°C **f** 54°C **g** 84°C **h** 76°C

3 Draw a ring around the temperature that is the odd one out.

35°C   15°C   64°F   21°C   48°C

Why is it the odd one out?

_____

4 This thermometer shows the temperature at
6 o'clock one winter morning in Paris.

The temperature rose 8 degrees by midday.

It rose another 4 degrees by 3 p.m.

By 5 p.m. it had dropped 6 degrees.

Draw a thermometer on a piece of paper to show
the temperature at 5 p.m.

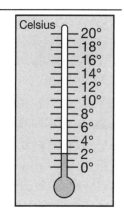

5 Would you prefer to visit a country where the temperatures are
about 30°C each day or about 2°C each day? Explain why.

_____

_____

Date: _____

Geometry and Measure

Lesson 1: **Position**

• Use language associated with position

1 Describe the position of the cube.

_____

_____

_____

2 Draw a triangle to the right of a square.

3 a What is the position of the sphere?

_____

_____

_____

b Where is the sphere now?

_____

_____

_____

4 Describe the position of the cube. Describe its position in three different ways.

_____

_____

_____

_____

_____

_____

_____

Geometry and Measure

5 Draw a circle, a square and a pentagon. The pentagon needs to be to the right of the square. The circle needs to be to the right of the pentagon.

6 a Where is the sphere? _____

b Where is the cylinder? _____

c Where is the pyramid? _____

7 Carlos has 3 shapes. He positions them in a line.
The blue cube is to the right of the sphere.
One of the shapes is red.
The yellow cylinder is between the sphere and the cube.

a What is the order of the shapes, from left to right?

_____

b Order the colours of the shapes, from right to left.

_____

8 Draw your own picture with five different 2D shapes. Describe where each shape is positioned in as many ways as you can.

_____
_____
_____
_____

Date: _____

Lesson 2: **Direction and movement**

- Use language associated with direction and movement

1 Circle the clockwise turns.

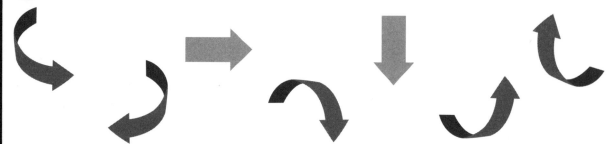

2 Circle the anticlockwise turns.

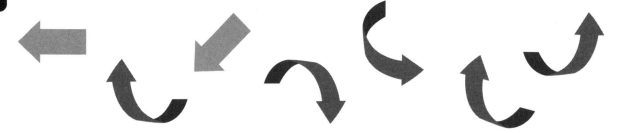

3 a You make a quarter turn anticlockwise. Then you make another quarter turn anticlockwise. What size turn have you made?

_____

b Where will you be facing if you make a whole anticlockwise turn?

_____

4 Draw a cross in the **top left square** and then follow the instructions.

a Draw a cross 4 squares to the right.

b Draw a cross 3 squares below your last cross.

c Draw a cross 2 squares to the left of your last cross.

d Draw a cross 1 square above your last cross.

e Draw a cross 2 squares to the left of your last cross.

f Draw a cross 2 squares below your last cross.

Geometry and Measure

Geometry and Measure

5 Draw a road sign to show a left turn.

6 Draw a road sign to show a right turn.

7 Sally says:

Do you agree with Sally? ☐

Why?

_____

_____

> If I make two quarter turns in a clockwise direction and then two quarter turns in an anticlockwise direction, I will end up where I started.

8 Draw 5 crosses on the grid. Join them with horizontal and vertical lines.

Use the words **left**, **right** and **up**, **down** to describe how to get from one cross to another.

_____

_____

_____

_____

_____

9 Write two sentences to describe movement. Use the word **clockwise** in the first one and the word **anticlockwise** in the second.

_____

_____

_____

_____

Date: _____

201

Geometry and Measure

Lesson 3: **Compass points**

• Use the four compass points to describe position, direction and movement

1 On a compass, what do these letters mean?

a S _____

b W _____

c N _____

d E _____

2 If you are facing north, in which two directions must you travel to get from Start to the Finish?

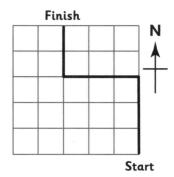

Finish

N

Start

3 **a** If you are facing north, what direction will you face if you make a quarter turn clockwise?

b If you are facing west, what direction will you face if you make a three-quarter turn clockwise?

c If you are facing south, what direction will you face it you make one whole turn?

4 Write the steps to get from Start to Finish.
The first step is written for you.

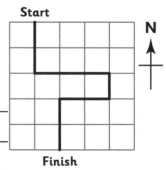

Start

N

Finish

Step 1: south for 2 squares

Step 2: _____

Step 3: _____

Step 4: _____

Step 5: _____

Geometry and Measure

5 Write the steps to get from Start to Finish.

The first step is written for you.

Step 1: south for 4 squares

Step 2: _____

Step 3: _____

Step 4: _____

Step 5: _____

6 Maddie is facing east. She makes a quarter turn clockwise. She then makes a three-quarter turn anticlockwise.

What direction is she facing now? _____

7 Compare these two grids.

a How are they the same?

_____

b How are they different?

_____

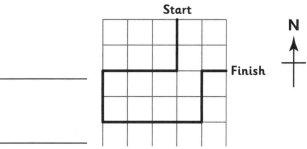

c Write the steps to get from Start to Finish on the first grid.

The first step is written for you.

Step 1: south for 2 squares

Step 2: _____

Step 3: _____

Step 5: _____

Step 4: _____

Step 6: _____

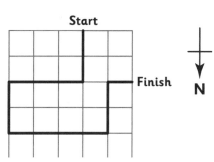

d Write the steps to get from Start to Finish on the second grid.

Step 1: north for 2 squares

Step 2: _____

Step 3: _____

Step 4: _____

Step 5: _____

Step 6: _____

Date: _____

Lesson 4: **Reflections**

Geometry and Measure

You will need
• ruler

• Sketch reflections of 2D shapes

1 Draw vertical mirror lines to show where each shape is reflected.

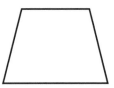

2 Draw horizontal mirror lines to show where each shape is reflected.

3 Here is a pentagon. One of its edges is along a vertical mirror line.

Draw its reflection.

What do you notice?

_____

4 Alice says:

Is this a good explanation?

Why?

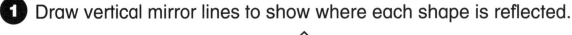
A reflection is a shape that has been copied.

_____

_____

5 Draw the reflection of each shape across the mirror line.

a

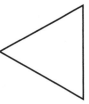

b

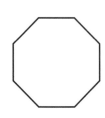

6 Draw the reflection of each shape across the mirror line.

a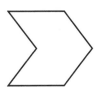

b

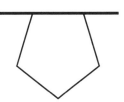

7 Which of the four reflections in **5** and **6** was the easier to draw?

Why?

_____

8 Reflect this pattern across the mirror line.

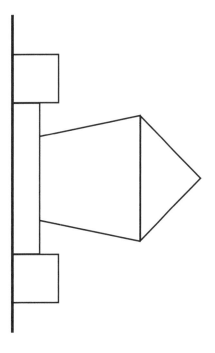

Geometry and Measure

Lesson 1: **Venn diagrams**

• Record, organise, represent and interpret data in Venn diagrams

1 Look at this Venn diagram. Write what is the same about cats and children. Write what is different.

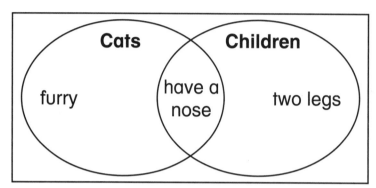

Tick the correct name for the box around this diagram:

Carroll diagram

Universal set

Frequency table

_____

_____

_____

2 Here are two Venn diagrams, each made from one set.

If we combine them, which numbers will go in the intersection?

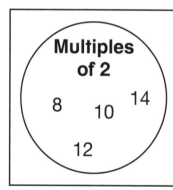

Multiples of 2

8   10   14

12

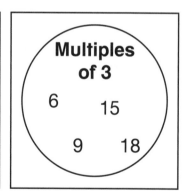

Multiples of 3

6   15

9   18

3 **a** Draw these shapes in the correct places on the Venn diagram.

circle

semicircle

triangle

square

pentagon

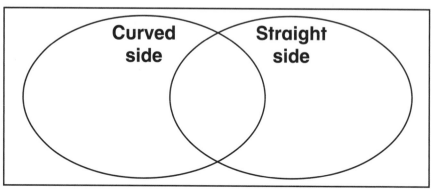

Curved side

Straight side

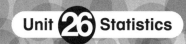
Statistics and Probability

b Write three statements about your Venn diagram.

_____

_____

_____

4 What could be the criteria for this Venn diagram?

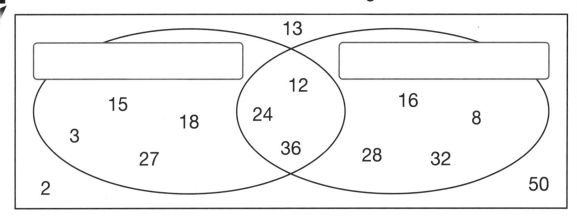

5 Mr Plum wants to add another fruit to those he sells in his shop. He doesn't know which fruit to choose. Can you help him? Choose two fruits. Ask the rest of the class the question 'Which of these fruits do you like?' Each learner chooses one fruit, both fruits or no fruits.

Add the data you collect to the Venn diagram. Don't forget to label the diagram.

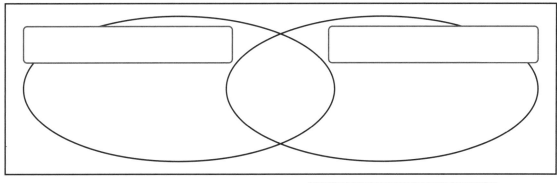

Which fruit should Mr Plum choose? [   ] Why?

_____

_____

_____

Date: _____

Lesson 2: **Carroll diagrams**

Statistics and Probability

• Record, organise, represent and interpret data in Carroll diagrams

You will need
• paper

1 Write or draw the shapes into the correct parts of this Carroll diagram.

circle

semicircle

triangle

square

rectangle

pentagon

hexagon

4 sides or fewer	Not 4 sides or fewer

2 Look at this Carroll diagram.

	Multiples of 2		Not multiples of 2	
Multiples of 3	12	24	21	15
Not multiples of 3	26	32	49	65

a Which numbers are **not** multiples of 2 or 3?

b Which numbers are multiples of both 2 and 3?

3 a Write the numbers on the Carroll diagram.

6   8   27   19   16   35   36   45

	Even number	Not even numer
Multiples of 3		
Not multiples of 3		

b Add three more numbers to each section.

Statistics and Probability

c Write three statements from your Carroll diagram.

_____

_____

_____

4 Benjamin drew a Carroll diagram. He could not understand why there were no numbers in two of the sections. Explain why this is.

	Odd	Not Odd
Multiple of 2		4, 10, 20, 36
Not a multiple of 2	5, 13, 27, 33	

_____

_____

5 Write the criteria on the Carroll diagram.

	56, 168	49, 84
	24, 72	23, 55

6 Write two more numbers in each of the four regions on the Carroll diagram in **5**.

7 Choose two sports. Ask the rest of the class the questions: 'Do you like …? What about …?' Each learner tells you if they like the sports or not. Add the data you collect to the Carroll diagram.

Write three statements from your results on separate paper.

Date: _____   ☺ ☺ ☹

209

Lesson 3: **The statistical cycle**

• Recording, organising, representing and interpreting data

Statistics and Probability

1 This tally chart shows learners' favourite colours.

Colour	Tally
yellow	||||| ||||| |||
blue	||||| ||||| ||||| |||||
red	||||| ||||| ||||| |||
brown	||||| |||

a Write how many learners like:

blue [ ]   red [ ]   yellow [ ]   brown [ ]

b How many learners took part in the vote? [ ]

2 This tally chart shows the number of different fruits sold on a stall.

Fruit	Tally
papaya	||||| ||||| ||||| |||
watermelon	||||| ||||| |||||
kiwi fruit	||||| ||||| ||||| ||||| ||||| |||||
pomegranate	||||| ||||| ||
mango	||||| ||||| ||||| ||||| ||||

a What are the three most popular fruits?

_____

b Which is the least popular fruit?

_____

c What is the total number of watermelon and

kiwi fruit sold? [ ]

3 Adnan says: I think 99 people took part in this vote about fruit.

Do you agree? ☐ Explain why.

_____

_____

4 Write three statements about the tally chart **2**.

Include the words 'sum', 'difference' and 'more'.

_____

_____

_____

5 Think of a question you would like to ask the class. It could be about the class's favourite animals, foods, sports. Write five options for learners to choose from. They can only have one choice. Complete the tally chart with the results.

	Tally

Write three statements from your results.

_____

_____

_____

Date: _____

Statistics and Probability

Lesson 4: **Frequency tables**

* Record, organise, represent and interpret data in frequency tables

You will need
* paper

1 This tally shows learners' favourite animals.

Favourite animal	Tally
monkey	|||| |||| ||||
elephant	|||| ||||	
lion	|||| |||| |||| ||||
giraffe	|||| ||||
zebra	|||| |||| |||| |||

a How many learners voted for zebras?

b How many more voted for lions than voted for monkeys?

c How many voted for elephants or giraffes?

2 Complete the frequency table to show the information from the tally chart in **1**.

Favourite animal	Frequency

3 This tally chart shows the favourite sports of some Stage 3 learners.

Sport	Tally
football	|||| |||| ||||	
basketball	|||| |||| ||
cycling	|||| |||| |||| ||||
gymnastics	|||| |||| |||| |||| ||||
swimming	|||| |||| |||| |||

a Represent this information in a frequency table.

b How many more learners voted for football than for basketball? ☐

c How many learners voted for swimming or gymnastics? ☐

d How many learners voted altogether? ☐

4 Chongan says:
> I think frequency tables are tables with numbers.

Is this a good explanation of a frequency table? ☐ Why?

_____

5 Think of a question you would like to ask the class. It could be about the class's favourite fruits, toys, places to visit. In the first column of the frequency table, write six options for learners to choose from. They can only have one choice. Complete the frequency table with the results.

	Tally	Frequency

Write three statements from your results on separate paper.

Date: _____

Statistics and Probability

Lesson 1: **Pictograms**

Statistics and Probability

* Record, organise, represent and interpret data in pictograms

You will need
* 2 cm squared paper
* ruler
* coloured pencils

1 This tally chart shows learners' favourite fruits.

Show this information as a pictogram. Use squared paper.

Choose your own symbol. Each symbol should represent 1 learner.

Favourite fruits	
mango	ⅢⅢ Ⅲ Ⅰ
apple	ⅢⅢ |||
kiwi	ⅢⅢ	
banana	ⅢⅢ ⅢⅢ

2 Use your pictogram to answer these questions.

a Which is the most popular fruit?

_____

b How many more people voted for mango than voted for kiwi?

_____

c How many people altogether voted for banana and apple?

_____

3 Look at the pictogram and answer the questions.

a How many more green sweets than brown

sweets are there? ☐

b How many orange and red

sweets altogether? ☐

c How many fewer purple sweets than yellow

sweets? ☐

d How many sweets

altogether? ☐

Colour	Number of sweets
green	●●●◖
orange	●●●●
blue	●●◖
pink	●●●
yellow	●●●●●◖
red	●●●●
purple	●●●◖
brown	●◖

Key ● = 2 sweets

Statistics and Probability

4 The table shows some learners' favourite wild animals.

Animal	Number of learners	Animal	Number of learners
elephant	12	buffalo	9
lion	14	lemur	15
gorilla	10	tiger	21

Draw a pictogram to show this information. Use squared paper. Decide on a symbol. Each symbol should represent 2 learners.

5 Write three statements from your pictogram.

_____

_____

_____

6 a The PE teacher wants to buy extra equipment for a sport. He doesn't know which sport to choose. Can you help him? Decide on five sports and write them in the frequency table. Read out each one and ask the class to show which is their favourite. Record the results in the frequency table.

Sport	Tally	Frequency

b Now draw a pictogram on squared paper to show the information. Decide on a symbol and whether one symbol will represent 1 or 2 learners.

c Use the information in your frequency table and pictogram to make up two questions. _____

_____

Which sport should the PE teacher choose? [   ]

Date: _____

Statistics and Probability

Lesson 2: **Bar charts**

- Record, organise, represent and interpret data in bar charts

You will need
- 2 cm squared paper
- ruler
- coloured pencils

1 This table shows learners' favourite sports. Show the information as a bar chart. Use squared paper. Mark the vertical axis in steps of 1.

Favourite sports	
Sport	**Number of learners**
football	12
basketball	8
swimming	6
gymnastics	5

2 Look at your bar chart from **1**. Write three pieces of information that your bar chart shows.

_____

_____

_____

3 Look at this bar chart, then answer the questions.

a How many more like cola than orange? ☐

b How many, in total, voted for lemonade or orange? ☐

c How many were asked altogether? ☐

Our favourite drinks

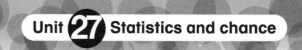

4 The table lists the musical instruments played by learners.

a Represent this information on a bar chart. Use squared paper. Mark the vertical axis in steps of 2.

b Beside your bar chart, write three statements about the data.

Instruments played	
Instrument	**Number of learners**
drum	20
piano	15
guitar	24
violin	10
recorder	9

5 Describe how a pictogram and bar chart are the same.

_____

_____

6 Describe how a pictogram and bar chart are different.

_____

_____

7 This frequency table shows the favourite sports of a group of learners.

a Represent this information in a pictogram. Use squared paper. Use one symbol to represent two learners.

b Now represent the information in a bar chart. Use squared paper. Mark the frequency axis in steps of 2.

Sports our class like	
Sport	**Frequency**
football	20
rugby	12
swimming	24
baseball	15
tennis	16

c Beside your pictogram and bar chart write five statements about the data.

Date: _____

Statistics and Probability

217

Statistics and Probability

Lesson 3: **Chance (1)**

• Use familiar language associated with chance to describe events

1 Tick the statements that **will** happen.

a There will be seven days next week. ☐

b Tomorrow will be yesterday. ☐

c It will snow next week. ☐

d The month after May will be June. ☐

2 How likely are these things to happen? Choose from these phrases:

will happen   might happen   will not happen

a Mum will cook tea tonight

☐

b An elephant will drive a bus.

☐

c I will play baseball for my school.

☐

d We will see an aeroplane at the airport.

☐

e Li will see a monkey eating bananas.

☐

f I will land on the Moon next week.

☐

Statistics and Probability

3 Choose the correct phrase for each sentence.

will not happen	might happen	will happen

a I will meet a dinosaur on my way home from school. _____

b I will eat some rice this week. _____

c I will be in school next Monday. _____

d Our TV will break down tonight. _____

e Our teacher will be up a tree reading a book this afternoon. _____

4 Explain your answers to **3** **a** and **b**.

a _____

b _____

5 Maisie says:

Do you agree?

▢ Why?

> We use 'will happen', 'might happen' and 'will not happen' to describe chance. October always comes after September. So that is something that will happen.

_____

_____

6 Make up your own statements to fit these phrases.

a Will not happen

_____

b Might happen

_____

c Will happen

_____

Date: _____

Lesson 4: **Chance (2)**

• Conduct chance experiments, and present and describe the results

1 Here are some digit cards. Write 'will happen', 'might happen' and 'will not happen'.

a I will pick a 3. [　　　　]

b I will pick a 2. [　　　　]

c I will not pick a 1. [　　　　]

d I will pick a 7. [　　　　]

e Explain your answer to **d**.

_____

_____

2 Jodie says:

Do you agree?

[　] Why?

> If I put these cubes in a bag and pick one, I am more likely to pick out a grey cube.

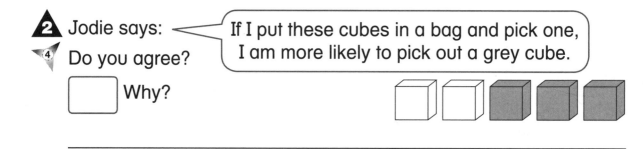

_____

_____

3 Look at the spinner. Do you agree with the statements? Write 'yes' or 'no' beside each one.

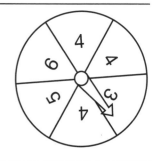

a I will spin a 4. [　] Why?

_____

_____

b I will not spin an odd number. [　] Why?

_____

_____

c   I will not spin a 3. ☐ Why?

_____

_____

d   I might spin a 4 or a 5. ☐ Why?

_____

_____

e   Write two other true statements for the spinner.

_____

_____

4 Rawal says:

Do you agree?

☐ Why?

> I have a blue cube, a brown cube and an orange cube. If I put them in a bag and pick one out, I will not pick a yellow cube.

_____

_____

5 Samir says:

Do you agree?

☐ Why?

> I am more likely to pick the number 12 from a pack of number cards to 20 than to get heads when I toss a coin.

_____

_____

6 Lauren says:

Do you agree?

☐ Why?

> I have this spinner. I think that I will spin a 6 more times than one of the other numbers.

_____

_____

Date: _____

Acknowledgements

Photo acknowledgements

Every effort has been made to trace copyright holders. Any omission will be rectified at the first opportunity.

p14t 1494/Shutterstock; p104 VectorOK/Shutterstock; p142 Siridhata/Shutterstock; p150l Sylfida/Shutterstock; p150c WEB-DESIGN/Shutterstock; p150r Lars Poyansky/Shutterstock; p164cr Stokkete/Shutterstock; p168bl Tropper2000/Shutterstock; p168bcl Mega Pixel/Shutterstock; p168bc Bogdan Ionescu/Shutterstock; p168bcr Dzm1try/Shutterstock; p168br Irin-k/Shutterstock; p173l Bogdan Ionescu/Shutterstock; p173cl Mega Pixel/Shutterstock; p173cr Dzm1try/Shutterstock; p173r Irin-k/Shutterstock; p184 Laurentiu Timplaru/Shutterstock; p218tl AnnaStills/Shutterstock; p218tc Digital Storm/Shutterstock; p218tr Suzanne Tucker/Shutterstock; p218bl Sorbis/Shutterstock; p218bc Svekrova Olga/Shutterstock; p218br Oleg Yakolev/Shutterstock.